SUPPLY CHAIN CYBERMASTERY

Supply chain cybermastery *is dedicated to those men and women around the world who display the courage to innovate and lead the creation of the new economy.*

SUPPLY CHAIN CYBERMASTERY

Building high performance supply chains of the future

Andrew J Berger and John L Gattorna

Gower

Published by
Gower Publishing Limited
Gower House
Croft Road
Aldershot
Hampshire GU11 3HR
England

Gower Publishing Company
131 Main Street
Burlington, VT 05401-5600, USA

Andrew J Berger and John L Gattorna have asserted their right under the Copyright, Designs and Patents Act 1988 to be identified as the authors of this work.

British Library Cataloguing in Publication Data
Berger, Andrew J.
   Supply chain CyberMastery : building high performance
   supply chains of the future
   1.Business logistics - Data processing 2. Business logistics
   - Computer network resources 3. Electronic commerce
I.Title II.Gattorna, John
658.7'0285

ISBN 0 566 08413 9

Library of Congress Cataloging-in-Publication Data
Berger, Andrew J.
   Supply chain cybermastery : building high performance supply chains of the
   future/Andrew J. Berger and John L. Gattorna.
      p. cm.
   ISBN 0-566-08413-9 (hardback)
   1. Business logistics--Automation. 2. Electronic commerce. I. Gattorna, John.
   II. Title.

HD38.5 B468 2001
658.5'00285--dc21   00-050293

Typeset in Plantin Light by Ian Kingston Editorial Services, Nottingham, and printed in Great Britain by MPG Books Ltd, Bodmin, Cornwall.

Contents

Acknowledgements

Writing this book has been a bold undertaking in the midst of the dynamic changes that are taking place in the world around us. To be anywhere near relevant, we needed to write and publish this work at speed, which meant getting a lot of help from friends and colleagues inside and outside Accenture. But of all who contributed, Linda Nuthall was first among equals. She has been the driving force throughout and to her go the accolades for a job well done. No doubt she will be relieved to get back to her real job as a manager in our Australian consulting practice.

Phil Churchman has been a source of enthusiasm and ever willing to explore new ideas. Dave Andrews has been ever ready with good-natured advice in between a busy travel schedule. Rob Woodstock and Paul Phythian contributed valuable insights into the evolution of eProcurement, and Jon Bumstead's grasp of eMarketplaces is unequalled – which is why he is leading our team at cpgmarkets.com in Europe.

Thanks also go to Mike Ethelston for his quiet and insightful support along the way, and Stella Coote for her patience and smooth aggression when things got difficult and the authors could not be found. Close colleagues Jamie Hintlian and Charles Findlay were very supportive in this whole endeavour.

Peter Branch brought valuable insights to the fore based on his years of experience, as did Mike Overs, Kate Adey, Kate Trice, and Rod Kay and the whole Ultemate team, who were an inspiration – just when we needed it. Fiona Gibson, supported by her colleagues Rachel Barere and Mark Klinge, singularly and together have been a constant source of creative enthusiasm – what a great marketing team to run with the book from here.

Sharon Dowsett helped do the research for 'Looking over the cyberhorizon', supported by Wanda Brown and her wonderful knowledge management team. Special thanks go to Mark Reynolds for his great content advice, and Nick Vaney for his contribution to our thinking in eProcurement. And the list goes on. Alex Milward, Stephen Varley, Craig Rea, Paul McGowan, Simon Longhurst, Dominic Palmer, Frank Nugent and Mike Salvino all made significant contributions – for which we are very grateful. Similarly, our thanks to Owen Griffiths, Benjamin Green, Philip Monks, Alastair Birt, Richard Turner, and Scott Egler (for his CRM input). Stefano Lorenzi, Andreas Schroeder and Vicky St John all helped in their own way.

Yet perhaps the greatest supporters of all, especially in the early days of development were Dwight Dowdell and Susie Ward who had faith in the project and found us the funds to do the work; we hope they will be well satisfied with the result ... and the dessert is still to come! Jonathan Whittaker helped steer us towards valuable sources of information and Carrie Cussack and Richard Krahn were there when we needed even more guidance and inspiration.

Susie Poole was boundless in her energy in the early stages of research, and at the business end of the project Carmel McCauley of Future Perfect Communications did a beautiful job of editing. Now it only remains for Julia Scott at Gower Publishing to work her magic and we have something that should absorb the busy executive for many hours.

Foreword

Several forces have coalesced in the last five years to fundamentally transform how businesses operate and enable the creation of powerful new business models. Some companies have foreseen these developments and positioned themselves to respond; some have been 'blind-sided', but have scrambled to adjust and compete effectively; and some companies have fallen (and others will fall) by the wayside and decline, unable to adjust and create the new business models required to compete in the new millennium.

These new forces include *globalization* of both markets and sources of supply; the rapid expansion of the *Internet* and assorted capabilities; the continued evolution of *new transportation services* that expand supply chain options; new tools to *integrate design and manufacturing*; and the development of a new class of software tools to both *monitor inventory and event performance* and provide workgroup and project management capabilities to create and execute a response. However, by far the most profound developments have been the new software tools and communications technologies that have enabled the *extended supply chain*.

The extended supply chain is the integration and active collaboration of operations of the different players in the channel. This includes creating visibility, real-time planning and coordinated execution of activities across channel partners, and creating a new architecture for the entire technology infrastructure to link the channel players. New tools in the areas of eProcurement, eAuctioning, eDesign, supply chain collaboration and synchronization, exchange platforms, real-time event monitoring, optimal pricing, and logistics coordination expand the operational and strategic options of players in the channel. Enormous value is created in the form of lower transaction costs, greater operating flexibility, enhanced asset productivity, faster lead-times, and better customer service and intimacy.

Some companies have embraced these opportunities with vigour, and, not surprisingly, we have seen a growing gap between leading and average performing companies in almost every industry. *Supply Chain Cybermastery* captures the key learnings from these leading companies. It outlines both *on what* and *how* to compete in the world of extended supply chains. It provides insights, through scores of examples, on eProcurement, eDesign and Manufacturing, eFulfillment, and eWorking and workforce development. It is a valuable guide for understanding and creating the new business models that will allow your company to compete in the new world.

William C. Copacino
Global Managing Partner – Supply Chain Practice
Accenture
Boston, 01/01/01

Introduction – the eSupply chain: the 'brain' of the firm

The business world is moving ever faster in this new age of the electronic economy, with an ever-increasing level of sophistication required in order to manage the realities of today's supply chain. The current situation is not unlike the challenges faced by a rally cross team: having to navigate at high speed through unknown terrain, one foot on the accelerator, the other poised above the brake, making split-second decisions with only partial knowledge of the road conditions ahead.

In this new operating environment, companies face a bewildering array of change opportunities and many difficult decisions about the allocation of resources. Executives are being required to make critical decisions in response to the challenges posed by the dotComs and eMarketplaces, the requirement for entirely new capabilities and knowledge, and the emergence of new operating models in the supply chain. All this is being played out in a marketplace characterized by a shortage of business and technology skills and an ever-increasing scarcity of talent. The war for talent is well under way, and traditional multinationals are generally the losers.

In the past, companies could improve their supply chain performance through what we describe as the Seven Principles of Supply Chain Management. These could easily be applied in search of operational excellence, and indeed they represent a straightforward agenda for change:

1. Re-segment customers based on their underlying needs.
2. Customize logistics networks to correspond to this new segmentation.
3. Integrate demand and supply planning.
4. Differentiate products closer to customers.
5. Source strategically.
6. Develop supply chain-wide technology.
7. Adopt supply chain-spanning performance metrics.

The seven principles were once sufficient to keep companies' supply chains competitive and in the game – up until the dawn of the new economy. Now, and in the future, it will be necessary to go beyond operational excellence and engage in behaviours that reshape the way business is performed. Rather than focusing on principles, the new supply chain approach will represent a fundamental cultural shift, permeating the breadth and depth of the enterprise. We have described this new direction as the Eight Cultures of Value Chain Competitiveness:

1. Operational excellence and continuous innovation.
2. Extending reach into customers and suppliers through Internet technologies.
3. Compressing the supply chain to eliminate waste in time and resources.
4. Creating market-level contingency to allow a flexible response to the unexpected.
5. Optimizing price to maximize value and performance in the supply chain.
6. Learning to operate in the new eMarketplaces.
7. Building new business models and relationships.
8. Supporting organizational change and performance.

The real power of the eight cultures comes from the discipline they impose, separately and in combination. For example, operational excellence combined with a deeper reach into customer and supplier bases, and supply chain compression combined with the ability to operate effectively in virtual marketplaces, are both formulas for success. The next generation of successful companies will have learned how to mould these eight cultures to achieve new levels of competitiveness previously thought impossible to attain. At this point the eEnabled supply chain will in effect be the 'brain' of the firm and its customers and suppliers – the commercial central nervous system of the combined enterprise.

So what will be different in this new world of eEnabled supply chains? Information technology will move from the domain of the IT specialist to that of the line executive. Outsourcing in its many forms will increase exponentially, and for the right reasons rather than simply expediency. Improved partnerships in the form of long-standing relationships will replace the 'zero sum' games evident today. Ruthless standardization of data and data disciplines will take place, and the global reach of the

Internet will manifest itself at the local level. Aggregation and scale will increase in and across most industries, but in some cases will not be sustainable. 'Best' companies will come to exploit the advantages that they and their channel partners enjoy over less well-organized competitors.

All these developments will place great strain on the interface between business and government, where there is increasing evidence of misalignment. Governance and tax implications are still unclear, as is the future role of government in regulating the newly emerging global business environment.

With so many areas of interest in such a state of flux as we embarked upon this book, we decided that it was premature to attempt a substantial organizing framework for the proposed content. Rather, we chose to present a series of points of view on the emerging topics of interest in order to inform, test, and challenge the reader. A brief explanation of each topic follows and should provide an early insight into our thinking.

The impact of eCommerce on supply chains (Chapter 1) sets out to explore the most confronting issues to emerge from eCommerce, specifically the impact on executives and management practices. We examine the challenges and opportunities surrounding the various traditional and emerging business types and take a more detailed look at the Eight Cultures of Value Chain Competitiveness.

Learning to compete as value chains (Chapter 2) looks at the changes we have witnessed in the competitive landscape, the forces driving these changes and the way leading companies have responded. Most of the attention in this chapter is focused on the new type of competitive assault made possible via the Internet – value chain competition. We consider some of the lessons on value chain competition that traditional organizations can learn from the new leaders, and offer insights into what constitutes success in this new competitive world.

Driving real value from eProcurement and strategic sourcing (Chapter 3) is one of the hottest topics in the book because it offers quick results. We observe that relatively few sustainable successes have been documented in achieving savings from procurement spend. Early strategic sourcing efforts led to disappointment, but eProcurement is now renewing interest in the potential to achieve significant savings.

Taken together, strategic sourcing overlaid by eProcurement processes and technology is a logical starting point for those venturing into eCommerce.

The chapter sets out to define the scope of eProcurement and how it fits with strategic sourcing. We examine the key elements of eProcurement and the relationships between them and suggest several imperatives for successful implementation of eProcurement ventures. In closing, we focus on the key lessons to be learned from eProcurement and strategic sourcing experiences.

The eFulfillment challenge – the Holy Grail of B2C and B2B eCommerce (Chapter 4) explores perhaps the most vexed area of eCommerce, physical fulfillment. Unfortunately, the great progress made in the electronic side of the supply chain has not been matched on the physical side. The chapter examines the vital role of fulfillment in the new world and the changing attitude towards this component of eBusiness since Internet trading started. The issues surrounding fulfillment and emerging business models and solutions are examined from the perspectives of both business-to-consumer (B2C) and business-to-business (B2B).

The eDesign and eManufacturing challenge (Chapter 5) addresses the impact of the Internet and Internet-enabled technologies on today's manufacturing environment. We look at the drivers of change in design and manufacturing, and review the range of strategies being adopted by leading organizations and the associated benefits of each.

Learning to synchronize supply chains through eMarketplaces (Chapter 6) looks at different forms of eMarketplaces and traces how they have evolved, focusing on the types of players, the moves they have made, and the challenges they faced along the way. We predict the likely eMarketplaces of the future and the likely drivers of success. The chapter also examines supply chain collaboration and synchronization and predicts a vision of eSynchronization through eMarketplaces. We conclude by highlighting the key challenges and lessons to be learned from eMarketplace-led synchronization.

New information technology architecture for supply chains (Chapter 7) begins with our predictions on the winners and losers in the applications software arena and discusses the hardware platforms of the future, the impact of netsourcing applications and the new models for outsourcing functionality. The chapter closes by charting the potential implications of these changes on supply chain management.

New ways to deliver eWorking and continuous innovation (Chapter 8) focuses on how the Internet, intranets and related technologies are being used to eliminate time wasted on non-value added activity and improve the quality of communications. The idea is to surround employees with relevant information and put in place processes and metrics that release the power of employees to perform at superior levels. The chapter looks at the use of employee-centric portals to enable eWorking and the standardization of processes and improvements in communication technologies. It also focuses on new approaches to knowledge management and the keys to achieving a culture of 'continuous innovation'.

Looking over the cyberhorizon (Chapter 9) scans the operating environment ahead for early indications of oncoming 'Exocets'. While many trends and predictions may fail to materialize, we argue that you will be better prepared if you systematically scan the medium- and long-term horizons and progressively become aware of the competitive environment – rather than be taken by surprise. This chapter looks at several trends that we expect to have the most impact on supply chain management and suggests how the Eight Cultures of Value Chain Competitiveness can help to deliver success in this future world. We also discuss inspirational leadership and the critical role it will play in enabling organizations to deliver required changes to supply chain management.

As the new economy grows and increases its spread, leaders and decision makers need to face the challenges with conviction and assertiveness; this new environment will be unforgiving for those who hesitate. We scan the present and future environments and provide these leaders with both 'food for thought' and a set of insights into the new electronic age. Our aim is to provoke focused thought, prompt smart action and ultimately assist organizations in navigating and even shaping the twists and turns in the road ahead.

1 The impact of eCommerce on supply chains

Introduction

eCommerce has already had a transformational impact on the way that most companies do business. Change has been both rapid and fundamental, affecting markets, communications, competition, leadership, partnerships and business governance. Above all, as the economic assumptions that formed the basis for success in the old world are being overturned, eCommerce is increasingly affecting the ways in which supply chains operate. Interaction and collaboration costs have been slashed. Physical assets are no longer the cornerstone of competitive success and value propositions. It no longer takes years of work and billions of dollars to establish a business with a global presence – a fact we have observed with the emergence of companies like Cisco, Amazon.com and Yahoo!.

Despite all this change, many traditional business success factors have remained constant. We are entering a new phase in which large multinationals are reasserting their authority over start-ups. Changes in B2C eCommerce have been fast, furious and unsettling. Change wrought through B2B eCommerce will be more aggressive and more definitive in its conclusions. Some companies are well prepared, others are awakening giants and many risk becoming the victims of the next wave. Regardless of the hype, traditional concepts such as liquidity, scalability and profitably will rule.

In this chapter we explore the impact of eCommerce and the ways in which organizations are responding. We consider four key questions that link eCommerce to supply chain performance:

- How has eCommerce changed the way companies do business?
- What key challenges and opportunities are companies facing?

- How is eCommerce changing the way executives manage their businesses?
- What new cultures of competition are required for eCommerce success?

New ways of delivering success

eCommerce has obviously had a significant impact on most businesses, but how has this change manifested itself, and is it sustainable? In the United States, B2C start-up companies stole a two-year advantage over many of the slower traditional multinationals. Although Europe was slow to react to B2C growth, the B2B segment, which represents as much as four times the value of B2C business, was rapidly embraced by multinationals in North America, Europe and Asia almost simultaneously. Multinationals initially trailed start-ups in development of new capabilities, but they have caught up quickly and now have the opportunity to select the best ideas and technology from the B2B revolution. Those few start-ups that possess high-quality business concepts and sufficient funding will survive. We are now entering a more 'considered' phase of eCommerce in which companies will focus on liquidity and scalability to achieve growth and profitability.

As eCommerce has evolved, companies have been changing the way they do business in several areas: through new products and channels to reinvent customer interactions, development of entirely new business models, new access to critical capabilities, and new methods for customer-supplier relationships based upon the relative power of each party.

eCommerce stages

Companies have been undergoing progressive stages of development in their eCommerce capabilities (see Figure 1.1). In the earliest stages of broadcast, interact and transact, the emphasis was on creating customer and supplier awareness. Most organizations have integrated these capabilities into how they conduct business. The next two stages, integrate and collaborate/synchronize, involve using technology to fundamentally change the way business is conducted. These last stages will have most impact on supply chains of major companies; those that can achieve liquidity and scale the fastest will be most successful.

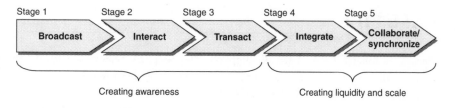

Figure 1.1 Five stages of Internet business.

- **Broadcast**: Companies started by investing in simple Web sites where visitors read material released by the company, such as corporate history, mission statements, product and service information, location and contact information. This remains a one-way broadcast channel.
- **Interact**: Web sites began incorporating two-way communication functionality, with features such as 'click here to send us your comments'. Customers who provided their email addresses did so on the understanding that, in theory, they would get their questions or comments answered or addressed. This was not always the case.
- **Transact**: Online selling emerged as early as 1996. Functionality included order taking, order confirmation and credit card processing.
- **Integrate**: Processes are automated and integrated, bringing real-time visibility of information to the customer.
- **Collaborate/Synchronize**: Virtual communities begin to take shape. Multi-dimensional communications occur online and in real-time to share knowledge, ideas, feedback and more. The old mantra of the systems integrators – 'simplify, automate, integrate' – has evolved into 'integrate, collaborate, synchronize'.

Although the stages emerged sequentially, companies today need not tramp the same path. First mover advantage in any one stage can be eroded by competitors that 'leap frog' to the next stages of development, especially when the competitor has a well-established brand. Success is not always being the first to move; it is more likely being the first to scale – after all, Amazon.com was the sixteenth to move but the first to scale!

New products and channels

New products and channels created by Internet technologies are a mixed blessing. They give companies, their competitors and new entrants expanded access to customers and can help them secure new customers through lower costs-to-serve. However, these benefits can be compromised by increased price and competition transparency. New

CASE FILE

Coca-Cola Amatil (CCA) Australia employs 3400 people, services 114 000 customers from national supermarkets to corner stores and produces 1450 million litres of beverages a year. Until June 2000, CCA used a paper-based system to transact with its trading partners. The company unsuccessfully tried to build its eCommerce infrastructure internally, and in early 2000, outsourced it to GE ECXpress, the Australian arm of General Electric's eCommerce division.

In three months CCA's rank jumped from the bottom 5 per cent to the top 5 per cent on retailers' SCOR cards. CCA now automatically processes large customer orders online within two hours, and is considering extending its online capabilities to its smaller retail customers. Before these reforms, retailers were threatening to remove CCA products from their shelves if it did not develop eCommerce capabilities.

channels allow commerce with customers across unlimited geography and time zones. Revenues may be lifted, but there are cost implications for managing round-the-clock operations. Disintermediation between traditional indirect channel partners can be another result of new channels, but may require the intermediaries to assume new service and support roles. For example, car manufacturers that begin selling direct to consumers will want car dealers to shift from a sales to a service focus, and chemical companies may push distributors to focus on logistics rather than sales. Similarly, online communities can secure more customers; however, a proliferation of communities increases the likelihood of switching and decreases its cost.

The opportunities to reduce costs by conducting business over the Internet are well documented. Savings have been realized by automating transactions, communicating electronically, and improving access to information that supports negotiation and strategic decision-making. The cost differential between traditional interactions with customers (such as sales force visits) and Internet transactions can be as much as 100 to 1. The allure of such savings has prompted some companies to re-evaluate sales operations and assets and make radical reductions in brokers, sales organizations and sales offices.

With new channels, there is also an increase in product syndication, or the sale of the same product to many customers without having to reproduce it in a physical sense. In the past, syndication existed only in the media and entertainment industries for software and music. However, with the Internet as the channel, many more products are suited to this practice. Intellectual property, technology and relationships are becoming re-sellable assets and comprising the core business for many eCommerce organizations.

The Internet itself has paved the way for some completely new products and services, such as online content management and portal services. At the same time, the Internet channel and associated new business models have greatly accelerated industry convergence; the potential range of product and service offerings for any given eTailer is far broader than that of traditional retailers through the erosion of traditional boundaries.

CASE FILE

Tesco, a leading British supermarket chain, has extended its online product range to include books, CDs and financial services. Future plans for Tesco even include cars. Tesco's fifty–fifty joint venture with The Royal Bank of Scotland offers Tesco-branded financial services such as home insurance, savings accounts and credit cards. It is envisaged that, in the near future, 45–50 per cent of their online revenue will come from products other than groceries. Similarly, Amazon.com has expanded its core business to include many products in addition to books – including such commerce businesses as their zShops programme. Small shops pay a listing fee to be hyperlinked to Amazon.com's Web site and receive a commission on sales conducted through the hyperlink. Amazon.com is now extending this programme to include larger stores, such as drugstore.com.

Tesco now considers Amazon.com to be one of its major competitors in the online grocery market, even over other traditional domestic food retail and supermarket chains.

New types of company

Developments in eCommerce highlight distinctions between high- and low-growth companies, leading to a significant shift of people and investments to those that are high-growth – a shift that increases the gaps between different types of company. We have identified four types of company based upon their capability to manage complexity and relationships in the eEconomy (see Figure 1.2):

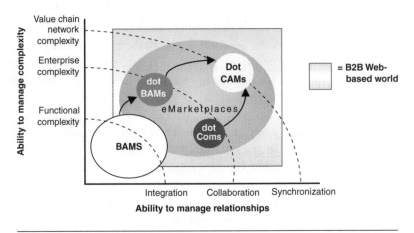

Figure 1.2 New types of company in the eEconomy.

- **dotCom start-ups** have focused a small number of talented people on good ideas. Most are more opportunistic than brilliant, and are characterized by the simplicity of their business models. They excel at managing relationships, but are not usually effective supply chain managers. They tend to develop good customer front-end technology, then make extensive use of people to compensate for supply chain inadequacies. Examples of dotCom start-ups include Amazon.com, ChemConnect and *simply*engineering. Growth rates are high, but the base is low.
- **BAMs**, or bricks-and-mortar companies, are traditional companies that earn good profits and have substantial revenues, but may suffer from fragmentation, inertia and an inability to engage in eCommerce. Their challenge is to retain talent and develop new competitive capabilities. Many have responded by spinning-off new dotCAMs; however, this is often a dangerous path. Such segmentation of capabilities can compromise the parent company's ability to react to eCommerce developments. Examples can be found amongst those companies that have fallen off the *Fortune* and FTSE 100 lists. Reflecting the maturity of BAM businesses, growth rates are low.
- **dotBAMs** are an interesting new breed of company that are growing out of the BAMs. They are applying the experiences of the dotComs and dotCAMs to their traditional organizations. This requires an ability to persuade large numbers of managers and employees to embrace eCommerce. Examples of dotBAMs are GE, Dow, Eastman Chemical Company and BP. These companies have learned to combine the weight of a traditional multinational with enthusiastic

eCommerce champions to develop strong links to the eCommerce world.

● **dotCAMs**, or clicks-and-mortar, are those rare organizations that have learned to combine an effective Web presence with an effective, responsive supply chain. This does not mean they own the key elements of the supply chain – most don't! However, they have learned to manage demand, supply and fulfillment better than most companies. Examples include Sun Microsystems, Intel, Microsoft and Cisco. Growth rates are high for these, even from a strong base.

The relationship between these different types of company raises several interesting challenges. Generally, the dotCAMS have been focused in narrow market areas. They will benefit from more companies, particularly multinationals, developing similar capabilities to create liquidity and scale. Many dotComs are struggling to become dotCAMs. It is one thing to set up an Internet-based customer interface, but supporting it with a world-class supply chain presents a more difficult challenge. DotBAMs are well-placed to reinvigorate growth in their organizations and take advantage of technologies that will be released from fading dotComs and dotCAMS. Those BAMs that are not well-placed for growth face an uncertain and difficult fate. As the gap between their existing capabilities and the new capabilities required by the eEconomy grows, they will find themselves increasingly isolated from growth opportunities and partnerships.

Building critical capabilities

Many organizations have already been on a long journey of efficiency-seeking initiatives that have resulted in the outsourcing of various components of their supply chains. Most of these companies will admit, although only behind closed doors, that they have lost the capability to reinvent many of the functions that they have neglected during years of cost-cutting. Today, organizations are improving their ability to understand where to focus their current internal capabilities and when to buy, build or borrow new capabilities.

● **Building** means developing capabilities internally. The key barrier is the time required to learn and implement.
● **Buying** means recruiting skilled employees or acquiring existing organizations that possess the required capabilities. The key barrier is generally cost.

- **Borrowing** means hiring, outsourcing or aligning with other organizations that possess the required capabilities. The primary concern associated with this option is the risk of loss of control.

CASE FILE

Cisco acquired more than 60 companies between 1984 and 2000. It has eight long-term alliances and is constantly developing new ones to meet customers' needs, ranging from technology development and solution implementation to customer support and new market entry. For Cisco, buying companies is an effective way of recruiting top talent. For example, Cisco gained 20–30 trained, experienced engineers from one acquisition.

The decision to build, buy or borrow capabilities should address a number of critical factors, including the quality and extent of existing capabilities, how easily they can be transferred to other areas in the business and the potential to convert them into new capabilities. The organization must assess how critical the control of customer contact, timing and interdependence with other capabilities is to the business. Consideration must also be given to the connectivity and integration between existing and potential operating partners, and the ease and expense of implementation.

In the old economy, 'hard' physical assets, such as facilities, equipment and other aspects of eFulfillment operations, while potentially expensive, were likely to be the simplest to acquire and operate internally. 'Soft' assets, such as information systems requiring internal and external integration, recruiting and training of the workforce, and alignment of leadership, culture, markets and execution processes, were the most difficult to perform well. Today, more and more companies are choosing to focus exclusively on the 'hard' functions, such as manufacturing or logistics (e.g. Solectron and Exel Logistics), while others are virtually asset-free and focus on information-based specialist activities such as flow management (e.g. Optimum Logistics and ShipChem).

There is a balance to be struck between integrating eCommerce into all aspects of an organization and treating it as just another channel. For most businesses, the answer lies somewhere in between.

CASE FILE

In 1999, KBToys.com, the Internet 'presence' of the traditional BAM, KB Toys, and an experienced dotCom organization, Brainplay.com, were merged to create KBKids.com. KB Toys determined that it had insufficient skills to create a full capability Internet business and sought a strategic alliance with Brainplay.com. The new entity sought to capitalize on the benefits of both integration and separation. The venture has two strong competitors, Toys "R" Us and eToys.

The merger integrated some aspects of the new business and segregated others. With a separate head office, KBKids.com's management team was made up of ex-employees from both original companies. Those from Brainplay.com had experience launching an Internet business and thrived in a fast-paced, entrepreneurial start-up culture. Three of the board members came from KB Toys or its then parent organization, Consolidated Stores. They brought valuable retail knowledge and were able to maintain visibility and control for KB Toys. In some operations, such as purchasing, KB Toys and Brainplay.com became fully integrated.

The KBKids.com brand, being similar to the KB Toys brand, leverages public awareness of KB Toys, but is distinct enough to ensure that the new company maintains its flexibility in pricing and product offerings. The variation in the brand name also allows the Internet store to extend its product offering beyond toys.

CASE FILE

Treating the Internet as just another channel to complement its mail order catalogue business, Office Depot determined that its existing distribution capabilities and information systems were appropriate for its Internet business. Office Depot already had existing catalogue sales operations such as taking and delivering orders to home addresses, and sophisticated information systems to integrate customer orders with real-time inventory data for 1825 stores and 30 warehouses. Additionally, its integrated eCommerce strategy is being combined with 3D interaction capabilities.

Changing power relationships

The evolution of eCommerce can also be tracked by its impact on buyer and supplier power. When buyer and supplier power are both high, there is an opportunity for mutual and direct partnership (see Figure 1.3). Where they are both low, channels such as distributors, agents and spot auctions are generally employed. Where there is an imbalance between

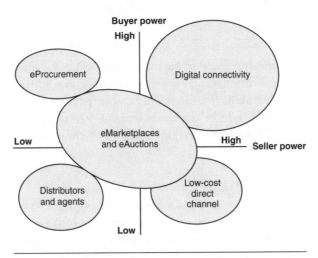

Figure 1.3 Customer–supplier power dependency map.

relative power, there are clear opportunities for eCommerce to change the dynamics of competition.

Low-cost direct channels were the first to embrace Internet technologies to service large numbers of small customers cost-effectively, recognizing the value of reducing cost-to-serve to increase margins. **eProcurement** was the second area to take off as companies with large indirect expenditure took advantage of new software packages to consolidate suppliers and product catalogues and to provide a friendlier user interface for buyers. **eMarketplaces and auctions** are emerging from both the buy- and sell-sides as companies seize opportunities to build platforms to connect organizations and transact in areas with little distinctive power in relationships. **Distributors and agents** are employed when there is little relative power between customers and suppliers. In such cases, Internet technologies are used to simplify business processes and provide larger companies with low-cost insight into the behaviour of marginal customers. The latest trend in eCommerce, however, is **digital connectivity**, the shift towards using Internet technologies to link high-value customers and suppliers. This includes the development of industry communication standards such as XML within vertical eMarketplaces. Although there are varying opinions about whether companies should manage high-value relationships through eMarketplaces, the key is to manage these relationships according to their degree of materiality, mutuality and maturity.

We have seen significant misjudgements as companies have experimented with customers and suppliers using inappropriate methods. Using eAuctioning against a long-term supplier, low-cost direct channels for high-value customers has not proven to be successful, nor has adopting high-value interactive approaches for low-value customers. The key to the power/dependence map is to recognize which strategies to use with which types of customer or supplier.

Key challenges and opportunities

With new opportunities come new challenges. Organizations must now manage increasing levels of complexity: extended hours; an increased workload in parts of the value chain by virtue of smaller, more frequent ordering and delivery; an expanded geographic base; shorter time frames; and a higher emphasis on time accuracy and excellence in service delivery. While many Internet start-ups have adequately mastered customer relationship management, most have yet to master product design and manufacturing, eProcurement, eFulfillment, demand and supply planning and after-sales service and support.

Developing ERP-linked business processes

An unexpected opportunity facing most companies is that of redesigning their core business processes to be smarter, more standardized, Internet-enabled and linked to traditional technologies such as ERP systems. Although some believe that Internet technologies can eliminate business processes, this is rarely the case. Procure-to-pay and order-to-cash processes must still occur – it is only the way in which they are performed, and by whom, that may change. The typical business (see Figure 1.4) is comprised of a number of standard processes (sell, buy, make and manage infrastructure). The impact of eCommerce can be viewed from an individual area such as eProcurement on the buy-side or Web-based selling on the sell-side as well as from the interaction between areas. These interactions can occur, for example, as design processes shared across R&D, manufacturing and marketing groups, as well as shared planning and forecasting between suppliers and customers and eFulfillment interactions between Internet customer services, logistics managers and third-party logistics providers.

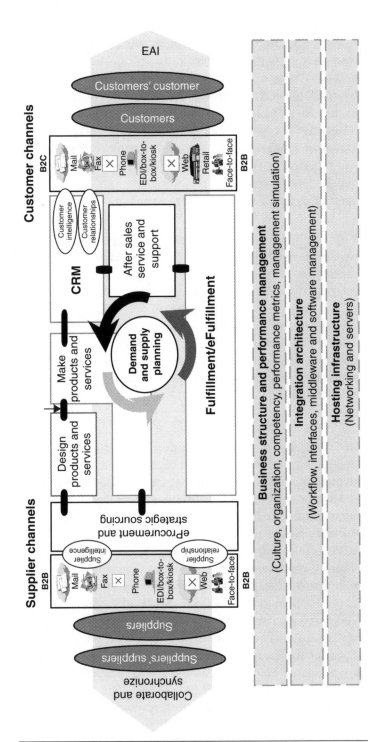

Figure 1.4 Supply chain interfaces.

One of the unique, and often unrecognised aspects of eCommerce, is the common standards it creates for communication between non-corporate entities such as government and academia. It is significant that the emphasis is on the ability to share data, using common language, standards and protocols, rather than on the ownership of the best proprietary standards. The Internet is similar to the air traffic control systems that control diverse aircraft and pilots from various nations by using English, setting flight paths, and establishing phrases such as 'mayday' as the standard rules for participation. In cyberspace, the Internet creates a system for interactions between companies and their competitors, partners, regulators, customers and suppliers. The success of these new standards varies by industry and depends upon how advanced EDI is in each industry. Regardless, we expect to see the rapid creation of intra- and inter-industry standards for all businesses.

CASE FILE

EAN Australia, a member of EAN International, has a membership base of approximately 11 000 companies and is responsible for coordinating, developing and promoting the EAN system of identification and barcoding in Australia. The EAN numbering and barcoding system identifies items, services, logistics units and locations by assigning a unique worldwide number to each. The system is administered in over 90 countries with approximately 850 000 user companies, and is used in all branches of commerce and industry worldwide.

EAN has also developed EANnet, a multi-industry electronic catalogue accessible via the Internet, EDI uploadable templates and custom software. It is aligned with UCCnet in the USA and other catalogues considered necessary platforms for emerging global eCommerce initiatives. Its aim is to ensure the integrity of data needed for eCommerce transactions by providing a single, standardized record of all required product data, including pricing and trading terms, that allow data and promotional synchronization between trading partners. EANnet also contains electronic images of products, primarily for shelf space management and the creation of virtual online shops.

Gaining speed, scale and sustainability

Key imperatives for success in the eEconomy are speed, scale and sustainability. Organizations must 'think big, start small and scale fast'. Strategic, collaborative alliances are now more important than ever. Companies are teaming to obtain skills rapidly, attain geographic reach and achieve critical mass. eCommerce is testing the limits of information technology functionality. The ability to capture and share Internet-enabled, real-time information has dramatically boosted the 'clockspeed' of many industries, dictating key elements of survival. 'Clockspeed', originally coined by Charles Fine in his book of the same name (Fine, C.H. (1999), *Clockspeed*, London: Little, Brown), refers to an industry's speed of evolution, manifested in areas such as the speed at which new products and services are introduced and new business models and strategies are adopted. Product and service life cycles are no longer predictable and long; consequently, strategies must be swiftly conceived, rapidly executed, and flexible enough to be changed at a moment's notice. eCommerce leaders are developing the capacity to experiment by employing strategic alliances and internal initiatives to better test strategic possibilities.

CASE FILE

The electronics and high-tech industry has experienced dramatic increases in its 'clockspeed'. The Internet has spurred a rush to introduce more products and product options to the marketplace in an effort to beat the competition. This speed-related competition has reduced the product life and profit cycles. Five years ago, for example, one could expect an average high-tech product to hold a viable market position for 18 to 24 months. Now, the product life cycle expectation is just 6 to 12 months, and even shorter for specialized products. Combined with an average market selling price decrease of approximately 1 per cent per week and steeper component price decrease, the product profit cycle is also significantly shorter today.

Many eCommerce experts have argued for the irrelevance of positive revenue and profit to eCommerce companies, focusing instead on a 'land-grab' for eCommerce territory and significance. This assertion may be correct for some niches. However, to achieve success, most companies must focus on a scalable, sustainable business model, building profit and continuous innovation around core products and processes.

Impact on management practices

Traditional effective management practices may not transfer directly to an eCommerce environment. Such an environment is volatile, wildly entrepreneurial and technically complex, generally calling for a new leadership style and corporate culture. Leadership approaches are shifting from leading people and functional processes to leading relationships in a fast-moving, performance-focused environment.

The electronics and high-tech industry is one that has the foundations and leadership dynamics that are most aligned with the characteristics of successful eEconomy leadership. Leaders in this industry have embraced a hands-off leadership style and succeeded in embedding the spirit of innovation, entrepreneurialism, calculated risk-taking and continuous innovation into their cultures.

CASE FILE

Nokia has been transformed from a Finnish company with a mixture of commodities and technology businesses into the world's leading mobile phone manufacturer and an Internet services provider. This has been achieved through CEO Jorma Ollila's creation of an entrepreneurial culture with a clear and focused strategy. Nokia's close and cohesive top management team takes a hands-off, consultative management approach in which decisions are made by those closest to the issue. This delegation of responsibility is said to encourage creativity and innovation. Control is maintained through results. For example, if a business line is not growing fast enough or does not demonstrate the potential to grow at a rate of 25 per cent per annum, it is not pursued.

One of the key lessons of the last two years is the need for companies to manage employees differently. For employees in traditional companies, flattened pyramids have meant fewer promotional opportunities. The challenge for these organizations is to spark growth and opportunity for their employees. Many companies have been slow to learn that traditional human resources systems, such as the Hay system, have failed to react to the increasing lack of upward mobility in flat organizations.

Additional management practices required by the eEconomy are the ability to manage complexity and the ability to manage relationships. The dotCom organizations tend to be very good at managing relationships, but not so good at managing complexity. They typically do a small number of things extremely well and use relationships to do everything else. The dotCAMs, such as Cisco, Intel and Sun, have good relationship management capability as well as the ability to manage a reasonably high level of complexity. Traditional BAMs are, for the most part, struggling to manage the relationships between their business units, let alone across companies. Many also struggle with complexity issues.

During the transition of their existing business units into eBusiness units, the dotBAMs, such as GE and BP, are improving their processes gradually in order to make them more relevant to the eEconomy. They are also making conscious decisions about core areas of focus and building stronger relationships with partners who can help them complete their full spectrum of capability requirements.

Connectivity between partners is the final essential element of new management practices. Without connectivity, there can be no effective use of electronic interfaces and no seamless information, money and physical product flows.

Eight cultures of value chain competitiveness

As we witness unprecedented growth in the eEconomy, we have seen the bar raised on best practices, changing the very definition of what constitutes a best practice. The following eight cultures of value chain competitiveness may serve as a high-level guide for successful organizations on the move.

Operational excellence and continuous innovation

In the eEconomy, people will want to partner only with those organizations that are operationally excellent and have a demonstrated culture of continuous improvement and innovation. Partners-of-choice will demonstrate these capabilities through their interaction with customers, suppliers and business partners. They will demonstrate a superior ability to exploit new ways to compete and avoid being challenged head-on by competitors.

Extending reach into customers and suppliers using Internet technologies

Successful organizations will have the capability to extend themselves into their customers' and suppliers' organizations using Internet technologies. Additionally, they will be able to reach through their customers to their customers' customers and up through their suppliers to their suppliers' suppliers.

Compressing the supply chain to eliminate waste in time and resources

Organizations, or more likely groups of organizations, will eventually improve the efficiency and competitiveness of their new business models by eliminating waste and time. Many will have fewer physical assets and fewer people. Leading organizations will learn to work with chosen partners to optimize the extended supply chain, leveraging demand and supply planning and decision support to manage a new style of network and its flows.

Creating market-level contingency to allow flexible response to the unexpected

Flexibility and agility must be achieved by planning for both the expected as well as the unexpected. Ultimately, leading organizations will operate without excessive slack while maintaining high levels of responsiveness. This concept requires the building of an industry-level shared and controlled 'safety stock' of reserves.

Optimizing pricing to maximize value and performance in the supply chain

Pricing of products and services has historically been a static feature of business, or at most has been approached from a product life cycle point of view. It is generally seen as an issue in isolation at the customer relationship end of the supply chain. A few clever organizations, such as Dell, have realized that pricing can be used as a tool to control demand and optimize flows and utilization across their supply chains. Dynamic pricing is a proactive approach to demand management. When there is pressure on supply, prices go up to stem the demand, and vice versa. This is particularly effective when one product can be played against or substituted for another. Dell uses dynamic pricing for its Pentium chips. For example, by adjusting its pricing of 500 MHz and 450 MHz Pentium chips every few days according to market activity, Dell is able to align component availability with demand. It is able to manipulate demand patterns to suit its long component part lead-times and short assembly lead-times.

Learning to operate in eMarketplaces

eMarketplaces will be a major feature in the business landscape of the future. Organizations must therefore learn how to interact with and operate in this arena. The ability to rapidly build new, key capabilities will be an entrenched skill and cultural characteristic of the leaders. They will have superior relationship management skills and performance measurement and monitoring mechanisms and will not be slow to sacrifice those who do not meet the required standards. To effectively operate in eMarketplaces, organizations must be able to manage high levels of complexity and be willing to share information and adopt agreed standards and protocols.

Building new business models and relationships

Successful organizations will use their relationships to gain access to complementary capabilities from other operationally excellent entities in their fields. They will extend their capabilities through alliances and partnerships without assuming additional assets or costs. There will be a saturation of the use of third-party and fourth-party logistics providers and the emergence of value chain networks.

Supporting your organizational change and performance

eEconomy cultures are not sustainable if an organization's people and culture are not properly aligned with the requirements of the new business models and strategic direction. In addition, like any high-performance sporting team, companies will need reserves. Employees will be expected to work longer and harder; however, they will need time to catch up and refuel. In the new eEconomy, operational excellence is critical, and to achieve high levels of performance, employees must be provided an environment in which to achieve sustainable results.

Conclusion

Before we leave this chapter, let us touch on a few final thoughts. A spate of regulatory issues are appearing on the horizon. Smart organizations are joining one or more of the emerging eMarketplaces to safeguard themselves against likely new 'anti-competitive' legislation. They are considering alternatives for the position or role they will play; many are looking at how they will use eFulfillment and other functionalities or services that potentially will be offered by horizontal business models. Of more certainty is the fact that matching demand in this economy will continue to be difficult and that getting the economics wrong will continue to be costly.

We should also expect to see many more hardware and software failures. The key reason for this is the lack of scalability being built into systems in favour of speed-to-market. Companies need to ensure that their new technical solutions and chosen platforms have the capability to scale. There are still big questions around whether certain platforms will scale beyond 25 000 users.

Companies must ensure that they have in place the appropriate processes and support mechanisms for both the new technologies they employ and the staff who are required to use them.

Expect waves of assault from new competitors. Companies must look for the gaps in their service and capability and have the capacity to fill them. There will be not just one competitive assault on businesses – attacks will come in waves as competitor entities extend their capabilities.

Finally, it is interesting to recognize that not all of the emerging business models are entirely new. In some cases, the business models being adopted are like those of years gone by. For example, new home delivery grocery models closely resemble the old world models in which a customer called the local store with an order, which would be home delivered by the 'delivery boy' after he finished his store duties for the day. In some ways we have come full circle, the difference being that we are now acting on a significantly larger scale, with powerful enabling technology and an extraordinary capacity for speed.

2 Learning to compete as value chains

Introduction

One of the key implications of emerging eCommerce business models is that the rules of competition are being transformed in ways never experienced before. Companies can no longer rely upon the same tried and tested formulae; in this new economy, there are new rules to be learned and new concepts to be experimented with. Some traditional key players have lost their dominant positions, while new entrants rise and fall rapidly. The smartest players will become the new aristocrats of leading stock market indices.

In this chapter we explore some distinct trends that are changing the terms of competition. We develop a picture of the evolution of value chain competition and discuss some of the lessons that traditional organizations can learn from the newer leading-edge companies – the 'clicks and mortar' or dotCAM organizations.

Our aim is not to advise traditional multinationals to copy these new business models – this would make little sense and would almost certainly destroy their unique value. Rather, our aim is to help traditional multinationals identify opportunities to adapt the experiences of the dotCAMs to their individual circumstances.

This chapter is focused on the following key questions:

- How is the competitive landscape changing, and why?
- What is value chain competition, and how will it create new value?
- What can traditional multinationals learn from the dotCAMs about value chain competition?
- What key capabilities will companies need to thrive in a world of value chain competition?

Changes in the competitive landscape

As companies increasingly see the possibilities and growth potential of value chain competition, we are witnessing an interesting evolution of competitive capabilities. The competitive landscape is shifting from intra-industry competition (The Best and The Rest) to inter-industry partnerships and alliances and partnerships of the top performers from different industries (The Best with The Best). As clear 'best partners' emerge, we predict a new 'eKeiretsu' style of value chain competition will emerge. Based on the industrial structure of post-war Japan, eKeiretsu competition will involve horizontally and vertically linked groupings of companies working with each other regularly.

The eKeiretsu groups will focus their time and attention on capturing new sources of value by competing as value chains against similar groupings of competitor companies. Such competition is not new – it has been evident in areas such as defence, construction and aerospace for many years. What will be different is the speed of decision making and execution enabled by new business models and Web-based technologies. Companies will be required to operate faster with more partners and in more complex areas than ever before – and then continuously accelerate their new capabilities.

'The Best and The Rest' – waves of competitive differentiation

For most businesses, competition has been fought out at a company versus company level within an individual industry: Coca Cola versus Pepsi, Cadbury versus Mars, American Airlines versus United Airlines, Visa versus MasterCard. The winner is generally the one with the strongest brand name and the ability to get its finished products to customers at the most competitive price. In these battles between industry players we have seen the clear emergence of industry leaders who over time stretch further and further ahead of their rivals. We have described this trend as being 'The Best and The Rest' (see Figure 2.1).

The leaders in this market structure possess several common traits. The traits most critical to their competitive positioning are the ability to develop new ways to add value and avoid head-to-head price competition by exploiting other ways to create competitive advantage. Their continuous evolution has allowed them to grow and innovate through what we call waves of competitive differentiation. The market leaders of

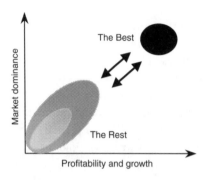

Figure 2.1 The Best and The Rest.

the last few years have recognized the advantage of assaulting their slower competitors with new capabilities while those very competitors are still reacting to the last wave of improvements. The aim is to convince the competition that they can never catch up.

Dell Computers illustrates this effect well. Dell has changed continually since its creation. It originally out-flanked Compaq, IBM and HP by its ability to sell directly to consumers. Dell's competitors failed to value this model, and even as it began to take hold they were hamstrung by their commitments to indirect channel partners, making it difficult for them to react. With the direct sales model, Dell created a new style of efficient supply chain management with such capabilities as make-to-order, postponement manufacturing and new approaches to supplier inventory, and gained competitive advantage through achieving high inventory turns and margins.

Once a new concept and capability has been introduced, such as Dell's direct model, it is only a matter of time before a competitor will replicate that capability and erode its differentiating value. There is little time to bask in the glory of a new idea. To stay ahead of the competition, the next ideas must be generated even while the new capability is being built. So while Dell's competitors sought to match the direct model – with varying degrees of success – Dell moved on to the next wave of competitive differentiation. The company has focused on relationship management capabilities and is venturing into new value-added service offerings such as *Premier Pages*, a tool created to help customers to more efficiently manage all phases of computer ownership, including purchasing, asset management and product support.

Dell's *Premier Pages* is a superior-quality service provided to key customers and designed to justify high market share and above average margins. *Premier Pages* provides customer organizations with a personalized Web page tailored to meet the specific needs. The page gives customers access to quality information and a fast and convenient way to conduct business. The service also yields significant savings for the customer organization through Dell's assistance in managing service and support. Dell has used its superior knowledge of customer buying behaviour to tailor its offerings to meet a demand long before any competitor senses a signal.

'The Best' working with new partners and alliances

The next phase in the evolution of the competitive landscape is organizations joining forces to gain competitive advantage. Under the new rules of competition, companies in the same industry – and sometimes even direct competitors – have started to cooperate while still competing with each other. They have been driven to collaborate by the need to scale and secure new levels of cost advantages. In the past, this cooperation was typically in the form of short- to medium-term alliances of convenience. However, some, such as the One World alliance and the Star Alliance in the airline industry, have become part of the fabric of their industries (see Figure 2.2).

Cross-industry alliances, such as Visa, Telstra and Qantas; BP and Safeway; Esso and Tesco; Somerfield and Elf; and Q8 and Budgens have started to emerge, adding another dimension to the competitive landscape. In such alliances, organizations are seeking to bridge the gap between themselves and the best in their industries by working with and learning from the best in other industries. This strategy is based on the premise that the characteristics that make an organization successful are

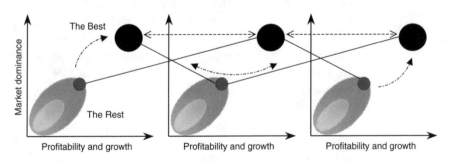

Figure 2.2 Learning to work with The Best.

transferable across industries; well-packaged, cross-industry service offerings can be very complementary to alliance organizations' value propositions.

Inevitably, many of these cross-industry alliance relationships will fail, but such failures will contribute to the learning process through which companies will discover how to select the best partners and effectively manage their alliances. British Airways, for example, struggled to build effective relationships with carriers in the USA such as United Airlines and US Airways before finally partnering with American Airlines.

The Best with The Best – supply chain versus supply chain

As companies have grown more comfortable working with partners from other industries, their focus has shifted to working with the best partners (see Figure 2.3). The key characteristics of being a best partner are being easy to do business with and possessing a strong focus on operational excellence for those processes that one partner needs from another. Partners in supply chains are learning to play a new competitive game of supply chain versus supply chain. This level of competition requires technology that spans the extended supply chain and facilitates end-to-end synchronization. It requires alliances with the best-in-class for all capability requirements, operational excellence and cross-functional integration at the individual company level, coordination, cooperation, data sharing, focus on end-to-end processes and elimination of sub-optimized processes.

Inevitably, the 'Best with the Best' model predicts a world in which the value of partnerships will be based upon the quality of the weakest links. Increasing value will be placed on working with those partners who can deliver on the promise of operational excellence and collaboration as

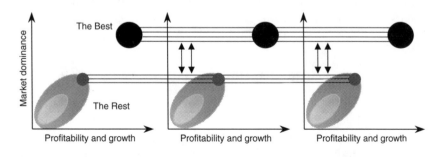

Figure 2.3 The Best with The Best.

CASE FILE

During the last decade, automotive manufacturers have streamlined processes with their first-tier suppliers by using EDI technology and forming collaborative relationships. The Internet is now enabling integration further up and down the supply chain so as to include the supplier's suppliers and customer's customers. Previously, this type of integration was prohibitively expensive due to the sheer number and spread of second-tier trading partners. DaimlerChrysler has capitalized on Internet-enabled capabilities and implemented several projects to make transactions seamless and transparent across the extended supply chain.

well as an ongoing agenda of continuous innovation. Early partnerships established between leading companies may leave latecomers with fewer options for successful partnerships. While the characteristics of a good partner will vary by value proposition, the key is understanding what capabilities are required for success and whether to build, buy or borrow the capability internally or externally.

CASE FILE

In October 2000, General Electric beat United Technologies Corp in its bid to acquire Honeywell. With a price tag of $42.9 billion, this is the largest and perhaps the most daring of the 1700 acquisitions that Jack Welch has overseen as CEO at GE. The three key reasons for the acquisition were a perfect fit between the two businesses, a common culture, and of course, the financial virtues of the deal. According to Welch, 90 per cent of Honeywell's operations fit with those of GE, and Honeywell's operations enhance four of GE's divisions. The greatest synergy is the strategic fit between the aerospace operations of the two companies, creating the dominant aerospace supplier in America and Europe. The new entity is capable of supplying everything from engines to electronics and maintenance services to jet makers such as Boeing Co. and the European consortium Airbus Industries. The combined aerospace businesses of General Electric and Honeywell will generate around $20 billion in annual revenue.

Value chain competition and the rise of the eKeiretsu

The essence of value chain competition is the combining of multiple partners to deliver unbeatable product and service offerings at a cost that is beyond the capability of competitors to match. In this next phase in the evolution of the eEconomy, companies are recognizing that it takes a great deal of time and effort to build such value chains. Value chain competition will be defined as multiple companies working seamlessly across multiple geographies and industries.

Given the opportunity to start afresh, few managers would ignore the accessibility to premium, lower-cost services offered by partners and instead opt to replicate these capabilities internally. Never before has there been such an opportunity for businesses to focus on their preferred core capabilities, rely on a network of others to provide non-core capabilities, and horizontally expand the parts of their business that produce the highest value.

In this new environment, certain companies will emerge as partners-of-choice in a given market. One of the key challenges for organizations is to select a position in the value chain network in which they can excel and to dominate that role or function. For example, UPS or FedEx in the fulfillment role or American Express in the billing and collection role. Organizations must consider which partnerships will provide the best opportunities for changing and improving the way their business is conducted. They must decide whether the opportunity provided by eCommerce is simply to have an impact the way information is exchanged or whether the delivery mechanism will be changed fundamentally as well.

CASE FILE

Cisco Systems, HP and EDS formed an alliance called the 'eCommerce dream team'. Together they provide their customers with a single turn-key Internet solution. Cisco provides the network technology and equipment; HP provides the enterprise solution, network operations and business support solutions; and EDS provides the business consulting and systems integration expertise.

Accenture (formerly known as Andersen Consulting) and Microsoft have aligned to form a new company – Avanade – that unites Microsoft's technology platform with Accenture's ability to deliver strategic business and Internet-based solutions to the marketplace. Avanade delivers Internet-specific and other enterprise platform services based on the Windows 2000 platform. In support, Accenture formed a Microsoft Solutions Organization, a cross-industry organization dedicated to architecting client business solutions based on the Microsoft enterprise platform. The venture will also focus on other joint programmes for marketing, business development and customer facing services.

eMarketplace and eProcurement solutions also provide a mechanism for partnering between organizations with no immediate and clear trading relationships, resulting in previously unseen power bases. Daewoo and J. Sainsbury, for example, piloted an alliance to create an alternative, non-dealer car distribution network. In addition to the creation of new delivery capabilities, many alliances are based on, or involve, a significant learning or knowledge transfer element. The alliance between Pillsbury and Webvan involves gathering intelligence about online consumer behaviours and effective methods for selling groceries and other consumer products over the Internet. DaimlerChrysler and Swatch's relationship is aimed at bringing trendy consumer design to the automotive business. BT, AT&T and Microsoft are learning about consumer and business usage of mobile Internet technologies, and SAP, BASF, Bayer, Siemens and others have joined forces to learn how to use an eMarketplace for collaboration, information exchange, and supply trading.

Similarly, major portal companies AOL and Yahoo! are creating new business models and value chains through their respective business alliances, mergers and acquisitions. For example, AOL has allied with General Motors and Yahoo! with Ford. Given that many consumers enter cyberspace through these portals, General Motors and Ford use the alliances to improve their reach to potential customers and improve relationships with existing customers. In parallel, the portal organizations increase their own usefulness, value and sustainability. This strategy has been adopted across a number of other industries – for example, Wal-Mart has teamed with AOL and Kmart with Yahoo!.

The aim of these relationships is to achieve a network of synchronized organizations that work together while maintaining a high level of agility. The challenge in developing an effective eKeiretsu group is the management of a complex portfolio of relationships. Assembling a network of best-of-breed allies who specialize and excel in the various links of the value chain can result in rapid growth and new levels of quality, flexibility and cost savings. To achieve this goal, companies must carefully choose partners who optimize market access and enhance the consumer experience.

eKeiretsu is a competitive concept that originates from the term Keiretsu, which describes the horizontally and vertically linked industrial structure of post-war Japan. The horizontally linked groups crossed a broad range of industries via banks and general trading firms and the vertically linked groups centred around parent companies with subsidiaries that commonly served as suppliers, distributors and retailers. Common characteristics among the groups included cross-holding of company shares, intra-group financing, joint investments, mutual appointment of officers, and other joint business activities. The Keiretsu system emphasized mutual cooperation and helped to protect its members from mergers and acquisitions. eKeiretsu describes a broader set of relationships that spans multiple companies and crosses national borders as a result of Web-based connectivity.

In an environment of value chain competition, companies will be challenged to create three key areas of capability:

- Successful building and managing of a portfolio of challenging relationships.
- Managing massive complexity and data across the value chain in real-time.
- Connecting the value chain through collaborative processes and technology.

These capabilities are discussed in the final section of this chapter and further expanded upon in subsequent chapters.

Learning about value chain competition from the dotCAMs

Many traditional organizations have struggled to appreciate the relevance of clicks and mortar companies. They are often heard to say that dotCAMs are 'new and different' and that comparisons between the dotCAMS and the bricks-and-mortar companies (BAMs) are not valid. However, the dotCAMs have developed a number of attributes relevant to traditional organizations competing in the eEconomy. Smart BAMs will inform the development of their own new business models and value chain networks with lessons learned from dotCAMs. In this section we introduce some of the areas in which the dotCAMs exhibit leading practices. We use a simple model for segmenting the enterprise: buy-side, sell-side, inside and outside the organization and connecting organizations.

On the buy-side

This is the area where most companies developing eCommerce strategies have 'cut their teeth'. The dotCAMs have developed procurement capabilities that avoid time-wasting, non-value-adding process steps. In addition to strategic sourcing disciplines, eProcurement and eAuctioning, leading dotCAMs have used capabilities such as eMarketplaces and procurement horizontals. Sun Microsystems, Cisco and Oracle are examples of companies that have achieved significant savings in procurement costs and avoided future costs as a result of Web-based technologies. This area of supply chain management has been confusing for even experienced players as a proliferation of software providers hold out the promise of massive benefits, easy implementations and new levels of supply chain connectivity. Chapter 3 explores this area in more detail and examines the key buy-side issues faced by today's organizations.

On the sell-side

The dotCAMs have also reached new levels of proficiency in their management of customer relations. Using the real-time connectivity afforded by the Internet and employing smarter ways to gather and employ customer data has enabled them to stay ahead of the marketplace. In many cases the dotCAMs are in fact the market-shapers, a phenomenon that could not have come about if their strategies and actions had been ill conceived or misinformed.

Electronic customer relationship management (eCRM) presents significant challenges. However, the dotCAMs have demonstrated the significant rewards that can flow from getting it right. Most traditional organizations have much to learn from the dotCAMs in the areas of customer service, attracting and retaining talent, motivating and rewarding talent, and building selling and service skills. Even call centre management capabilities are changing to align with the emerging cultures and behaviours of the eEconomy.

Inside the organization

The dotCAMs are working internally in new ways. 'eWorking' is a term we use to describe the ways in which leading organizations are leveraging the Internet, intranets and related technologies to eliminate time wasted on non-value-added activities. These companies are using employee portals and self-service mechanisms for an increasing range of administrative tasks and are standardizing processes and adopting new approaches for knowledge management so as to supply employees with the latest and most relevant information. Communication and collaboration tools and processes such as Webcasting and eRooms also are widely used by today's leading organizations. The dotCAMs have put in place processes and metrics to improve employee performance. The capabilities evolving in this area will help organizations to facilitate their change journeys into the eEconomy. As the success of all change initiatives relies upon the degree of support and cultural acceptance within organizations, we have dedicated Chapter 7 to this subject area.

Outside the organization

One of the most notable trends in the eEconomy is the increasing number of business connections being formed and the resulting explosion of new business models. Many of these new business models are being been driven by external investments in start-ups (e.g. Eastman and Dow Chemical investing in ChemConnect, Ford and General Motors in Covisint, and Dixons in Freeserve). The dotCAM organizations quickly learned that success would be difficult if they tried to do everything in-house. They knew that they could not grow fast enough or recruit enough talented people to succeed as independent entities in the eEconomy. These issues now are more widely recognized by large multinationals, and many are dismantling their operations in order to focus on their core capability, become a specialist in that field and use new partnerships to deliver all other required capabilities.

An increasingly common phenomenon that illustrates this concept is the utilization of outsourcing arrangements. Some companies never actually come into contact with their physical order. Instead, they accept most of their orders online and have relationships with contract manufacturers, contract administration service providers and contract logistics providers who execute their particular role in the supply chain with superior capability. Many also outsource the design process and online testing of products. High levels of automation and eWorking are characteristic of leading dotCAM value chain networks.

The dotCAMs are in constant search for new investments and do not constrain themselves to prospects within either their industry or the traditional supply chain structure. Traditional organizations can learn from the dotCAMs in their efforts to develop and ingrain a new culture

CASE FILE

65 per cent of items ordered from Cisco are not made or even handled by the organization itself. Orders are accepted through its Web page and immediately transmitted to a carefully selected supplier or manufacturer. The supplier has the capability to accept the order, arrange delivery through one of the carefully selected carriers, and even brand the product a Cisco product, while Cisco earns the greatest margin on the product.

Likewise, Nike uses contract manufacturers for all of its manufacturing operations and contract logistics service providers to manage its North American distribution centres. Honda too outsources approximately 75 per cent of its manufacturing.

that encourages their employees to be on the lookout for a new breed of opportunities, ventures and investments.

Connecting organizations

Clearly the dotCAMs could not have achieved their success without their initial commitment to laying solid, flexible and scalable foundations. They recognized that the key enablers in the eEconomy are streamlined connectivity and standardization. Traditional organizations are well advised to shift from the all too common fragmentation of systems and instead adopt a single ERP or 'ERP Lite' system as a foundation from which to access additional functionality from eCommerce and bolt-on application packages, integrated by middleware systems. With the foundations in place, the dotCAMs have developed the common processes, languages and databases that allow them to capitalize on the investment.

Connectivity between members of eKeiretsu groups and value chains is the other essential technology dimension in the new competitive environment. Leading dotCAMs have streamlined processes across partnering organizations and established common languages and protocols so as to capture the significant benefits of eSynchronization.

Thriving in a world of value chain competition

Value chain competition will challenge traditional organizations and punish those companies that are slow to react. The new capabilities required of companies are generally extensions of those activities they already perform. They will be required to do what they do better, faster and more consistently. And, they will need to coordinate multiple activities just as speedily and consistently. The dotCAM companies have demonstrated a path forward in terms of their capability to manage operational excellence, evolving customer needs, relationships, network complexity, shared standards and performance management in parallel. The challenge for more traditional companies is to orient their larger, more diverse organizations around simpler, more standardized methods of operation. It is a strange contradiction that the capability to manage extraordinary speed and complexity seems to have its roots in simplicity. The old message of KISS – keep it simple, stupid – seems as relevant today as it was in the past.

A solid logic based on operational excellence

A value chain network must make sense. This means assembling the right portfolio of partners, not just partnerships for partnership's sake. The pressure that this environment creates, and its heavy emphasis on first mover advantage, can prompt executives to rush precipitously into alliance deals. While swift founding of alliances is critical, companies must not abandon the fundamental principles of partnering. Alliances

CASE FILE

Cisco forms strategic partnerships with leading companies who have complementary technologies and services. It forms alliances to meet customers' needs, including offering end-to-end eCommerce solutions. Relationships are founded for joint development of technology, solution design, implementation and solution support, and are all based on operational excellence. These collaborative partnerships help to optimize the performance of Cisco's products and provide increasing value to the customers of all parties.

Cisco has carved a strong position as the market leader in network infrastructure technology and equipment. In 2000 it ruled 60 per cent of the router market, the company's original product, and more than 75 per cent of all Internet traffic travelled over Cisco products. To help maintain its leading position, Cisco manages a number of long-term technology and marketing alliances and constantly seeks and forms others when appropriate. Cisco forms relationships – such as the one with IBM – that involve sharing intellectual property and customer information to meet customers' technological requirements and establish a strong position for itself in anticipation of future trends – which Cisco and its partners are actively influencing.

must have solid, logical foundations and be operationally excellent; all partners must contribute something of enduring value, not just, for example, an exciting new application that may be obsolete tomorrow. It is critical to be wise and selective in deal making.

A true understanding of end customer needs

Most companies under-leverage their customer base, but value chain network winners will take the outside-in view. Customer equity is a new measure of success and is determined by taking the aggregate value of the share of market, the share of customer spend, and the customer lifetime value. Customer segmentation plays a critical role in eCommerce by enhancing the consumer experience and assisting in the proactive management of demand channelled through the Internet. A lack of customer segmentation has already caught many eBusinesses unprepared.

Value chain network competition requires identifying the range of buying behaviours that exist in the extended marketplace and the customer segments that are most attractive to the network in terms of their strategy and market positioning – that is, the most profitable segments. The organization can then develop market offerings and value propositions that appeal specifically to these target segments and explore the effects and possibilities of product features and price. The smartest networks will address issues such as dynamic pricing and will monitor the effects of pricing on volume.

Segmenting and targeting customers is a mechanism for competitive differentiation. By doing so, the value chain network will be better equipped for appropriate service provision and elimination of over- and under-servicing.

Ability to create and manage a portfolio of valuable relationships

As an exercise, write down the names of the three companies operating in your end-to-end value chain that you would least like to see working together due to the power that they could wield. This type of arrangement begins to resemble a value chain network. Think also about the three companies that could help your company to be unbeatable in the marketplace.

Your Worst Nightmare _____ + _____ + _____

Your Dream Team _____ + _____ + _____

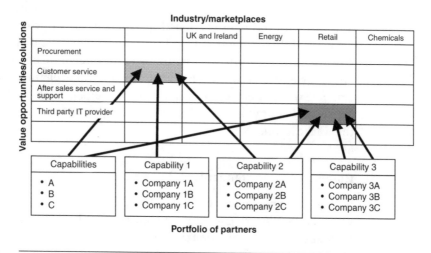

Figure 2.4 Working with a portfolio of partners.

The ability to manage relationships is greatly enhanced by the willingness of all players in the value chain network to share real-time information with all other players. Successful value chain networks will consist of a portfolio of partners who not only understand the end-use of their products and the behaviours and segmentation of their customers, but also understand each other and the joint and independent agendas of all participants (see Figure 2.4). The winning value chain networks will understand what they must do to respond in the marketplace and the role that each participant plays in the overall delivery. A level of trust must exist between value chain network partners such that data, information and knowledge can move throughout the network with efficiency, openness and accuracy; we call this 'transparency'.

Ability to manage complexity

Value chain competition requires each partner to have the ability to manage and collaborate around the complexity of the supply chain. eCommerce not only allows multiple companies to connect in a network, but also provides the flexibility for companies to join or leave the network on short notice and for the products and services to change and develop quickly. In this sense, the business conducted within the network will be more dynamic and complex than earlier generations of systems allowed. The combination of integration and the flexibility to change business relationships quickly creates opportunity.

Success, however, requires that all partners in the value chain network have the ability to manage high levels of complexity and change and that they all have the ability and willingness to collaborate on decisions that affect the entire network. The capability of a value chain network to compete will not be determined by the participant with the best capability, but by the one with the least. Buffers should be built around the weakest link.

Common shared standards

Successful management of the relationships and the complexities of value chain network competition requires a high degree of connectivity, trust and cross-organizational support. Speed and accuracy of information transfer is increasingly critical and is achieved through the establishment of shared language, standards and protocols. Every member of the value chain must be committed to achieving the same levels of operational excellence and continuous improvement. Each must bring measurable value to the competitive entity. Creation of buffers, or value chain redundancies, around the weakest links is an approach best adopted only for an acceptable and agreed period of time during which the strong players support and lift the weaker ones to the capability level and standard of the rest. To speak the same language, and avoid error and wasted time and effort through translation and interpretation, is as vital now as finding the right answer.

Performance monitoring and flexibility

Finally, value chain competition requires a willingness to acquire or drop partners on the basis of performance. This necessitates an effective performance measurement and monitoring system. The system should not only report outcomes, but should also guide improvements. In the fast moving eEconomy, more so than ever before, an effective performance monitoring system must be both flexible and responsive.

Conclusion

One of the standout features of the eEconomy is the fact that a critical mass of outsourcing options has been achieved. The implication is that vertical integration as a business model no longer makes the most sense. Instead, the best structure for winning in the eEconomy will be a horizontal and networked one. Instead of two-way trading relationships between a supplier and a customer, business will increasingly revolve around value chain networks made up of portfolios of partnerships, both natural and unnatural, between companies both within and across industries.

Earlier in this chapter we introduced the concept of 'waves of competitive differentiation'. Today, only discrete companies still adopt this strategy. We predict, however, that as they mature, the first round survivors of value chain network competition will also adopt this strategic approach. That is, a value chain network, as a unified entity, will endeavour to embed the entrepreneurial, innovative mind-set into the network's culture in order to ensure that it continuously redesigns and refreshes its competitive edge.

3 Driving real value from eProcurement and strategic sourcing

Introduction

The rise of new eProcurement solutions has captured the interest of boardroom executives in nearly every industry. With increased awareness of both the sizeable benefits and the implementation challenges, there has been a blizzard of daily press announcements reflecting the fast-moving and multi-layered marketplace for eProcurement solutions. eProcurement has become the starting point for many companies' overall eCommerce strategy, and many early adopters, declaring eProcurement success (and even victory!) have expanded their efforts into the emerging B2B eMarketplaces. Amidst the rush to adopt leading-edge procurement technologies there are some fundamental questions to be answered:

- What is eProcurement, and how does it fit with strategic sourcing?
- How do the key elements of eProcurement fit together?
- Where is the demonstrated value, and what are the critical drivers of successful approaches?
- What are the key lessons to be learned from eProcurement and strategic sourcing experiences?

Procurement spend, representing between 40 and 80 per cent of cost of goods sold in many large organizations, has long been a key element for management attention. Although it is an obvious area for cost reduction, our experience shows that only those companies that take a strategic approach to procurement and the role of eProcurement can successfully deliver the expected benefits. The best programmes are owned by a strong procurement organization, known for good customer service, which dedicates sufficient resources to strategic sourcing and progressive supply management, and engages the wider organization to meet the eProcurement challenge. The best programmes generate impressive initial results, putting structures in place to ensure that corporate contracts remain competitive and that benefits are sustained.

This chapter provides perspectives on eProcurement and strategic sourcing by relating eProcurement to the fundamentals of delivering value from procurement.

eProcurement and strategic sourcing

eProcurement is the value-added application of eCommerce solutions to facilitate, integrate and streamline the entire procurement process, from consumer to supplier and back again.

eProcurement initially became synonymous with the automation of order placement transactions, often referred to as eRequisitioning or eBuying. eRequisitioning enables users within an organization to use a simple Web browser tool to place orders with suppliers from an electronic catalogue of pre-negotiated items. eRequisitioning applications have been used by many companies to introduce a common, simplified system-based purchasing process to reduce the maverick buying that is prevalent in indirect procurement spend areas.

More recently, the scope of eProcurement has expanded to cover the full remit of processes and skills that justify its classification as a value-added application for the entire procurement process. eProcurement now incorporates eContracting and eIntelligence, while eRequisitioning is expanding into the direct area of business spending.

Strategic sourcing is the identification of a strategic mix of suppliers to fulfil an organization's commodity and services demand with improved cost and service. Traditionally the benefits from strategic sourcing programmes have not been sustained because of lack of compliance. eProcurement now offers the compliance mechanism through which the value delivered by strategic sourcing initiatives can be locked-in by linking strategic sourcing programmes to an eProcurement solution.

The use of a single system-based procurement process, providing wide access to corporate contracts developed as part of a strategic sourcing process, improves an organization's ability to manage compliance to the corporation's best contracts, and therefore maximize benefit delivery. It also provides improved spending data to those managing the relationship with suppliers, which can be used for fact-based negotiation of future contracts, thereby providing further improved costs and services.

Moreover, those companies that maintain strong relationships with suppliers are likely to drive most value from eProcurement. The improved ability of those within a company to surf the Web and identify new suppliers with seemingly cheaper prices does not guarantee the level of service delivery performance, the longer-term operational performance of the finished product or the total cost of ownership that can be derived from strategic sourcing arrangements. Strategic sourcing is also able to take into account the variety of product-level and service-level variations from the same supplier that may be required in an organization.

When combined with strategic sourcing, eProcurement can reduce the supply base, achieve significant process improvements and create a new platform from which to deliver improved purchase performance. The outputs from an eProcurement solution, specifically in terms of performance reporting and spend analysis, can significantly improve the information and decision-base of subsequent sourcing exercises, each part effectively closing the loop of well-rounded purchasing performance.

The best eProcurement approaches from companies account for several complex factors:

● *The eProcurement solution marketplace is fragmented, complex and immature*
  The eProcurement marketplace is comprised of more than 100 applications, mainly focused on a single building block – eRequisitioning. A variety of service providers, such as third parties that convert, aggregate and host electronic catalogue content or Application Service Providers (ASPs) that host eProcurement tools, add complexity to selection decisions. Except for a handful of companies, notably Ariba, Commerce One, Netscape, SAP, Oracle, and Metiom, few solution providers can demonstrate a strong track record or point to meaningful reference sites. Early leaders have tried to consolidate the maturing market by acquiring software to broaden their offerings beyond the eRequisitioning building block. For example, Ariba acquired Trading Dynamics and TRADEX Technologies, while Commerce One acquired CommerceBid to add eMarketplace and auction solutions to their offerings.

The best companies manage this complexity by developing content and technical strategies that enhance the solutions offered and protect them from the extreme case if any of the service providers ultimately fail.

- *The market has been focused on software-led approaches*
  The best eProcurement programmes are driven by holistic capability building decisions rather than just by software selection. This approach encourages starting with a strategic sourcing review and addresses implementation challenges such as large-scale content rationalization, process simplification and Enterprise Resource Planning (ERP) integration up-front. eProcurement benefits are not delivered through implementation of a tool; rather, real value is derived from the development of long-term online capabilities. These companies focus on the true drivers of best practices in procurement capabilities.
- *Procurement intelligence tools add substantial value*
  Procurement is ultimately a knowledge and intelligence game. Companies that recognize the link between a successful procurement solution and accurate, meaningful data are the ones that are able to justify investments in eIntelligence and thereby enhance the value derived from strategic sourcing.

Elements of an eProcurement solution

Our view of the eProcurement landscape is broader than the typical focus on eRequisitioning. Our goal is to increase a company's ability to tie fact-based negotiation to execution of contracts. It includes three distinct building blocks, each of which addresses different spend areas to deliver different levels of potential savings (see Figure 3.1).

- **eRequisitioning** has, until recently, been focused on indirect goods and services. Process savings have been achieved through eliminating inefficiencies in accounts payable and the 'lock-in' of negotiated prices from corporate contracts. Companies have targeted about 5–8 per cent in spend savings addressed by eRequisitioning.
- **eContracting** has the potential to maximize the savings negotiated in these corporate contracts by applying dynamic pricing techniques such as reverse auctions to direct and indirect spend categories. In our experience, eContracting has delivered between 6 and 35 per cent savings on current contracts.

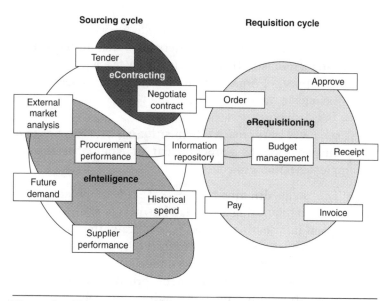

Figure 3.1 The eProcurement solution space.

- **eIntelligence** provides timely and accurate data from internal and external sources to enable the implementation of the best supply arrangements across all areas of external spend. The most successful negotiations are dependent upon an in-depth understanding of spend data, product and service characteristics and the supply market.

Winning companies develop an integrated set of capabilities. There is a tremendous opportunity for those companies that can unite all three building blocks under one coherent strategy.

eRequisitioning

eRequisitioning has introduced self-service purchasing to casual buyers in large organizations. The unique proposition of eRequisitioning is that it enables users to select pre-specified items from an electronic catalogue at better contracted rates than individuals could attain, and to process those orders electronically with suppliers in as efficient and cost-effective manner as possible. As one purchasing director from an international travel company said to her department at a presentation to communicate the company's eProcurement vision:

'... a browser-based user interface and electronic content will ensure that the deals we put in place are used as they were meant to be used.

Workflow will enable our people to use simplified processes (relieving us of paperwork) and suppliers to have more accurate real-time interaction with them. ... Put simply, we have the potential to extend the appeal of Amazon.com throughout this business whilst delivering savings to the bottom line.'

A good way to approach eRequisitioning is to review the major challenges faced by a procurement director embarking on an implementation project:

- The eRequisitioning business case is conceptually easy to understand. However, building a detailed business case for indirect spend, where spend information is poor, is less straightforward. As with any procurement initiative, the effort required to track the benefits to the bottom line is often underestimated.
- Using eRequisitioning to enable broader reengineering of the requisition-to-pay process requires close coordination between stakeholders from the procurement, finance, information technology and audit departments. Changes made to receiving and invoice-matching processes, for example, will drive the definition of interfaces between the eRequisitioning tool and a company's legacy business systems. This coordination across departments, though complex, ensures the practical simplification of historical processes.
- Electronic catalogue content must be user-friendly, contain accurate item descriptions and prices, and be translated into meaningful order information for suppliers. There must also be consistency of commodity/service code mapping in order to enable meaningful spend analysis. The skills required to create and maintain electronic catalogues do not match the traditional skills found in most procurement departments. Suppliers are similarly challenged to create electronic catalogue data, especially when approached by multiple buying organizations with varying information needs. Third parties provide content conversion services; however, the buyer or supplier must pay on a one-off or pay-per-use basis.
- Users will only accept and use eRequisitioning tools if they are easier to use than existing processes and if there is a penalty for not using them. Many organizations have struggled to create a positive environment for user acceptance of eProcurement solutions – the best answer is to create these conditions. Few users means few transactions and therefore few benefits.

Global 2000 companies have the greatest opportunity to drive consistency and leverage spend through eRequisitioning because of historical fragmentation. However, the challenges are multiplied for global, multi-business companies. For example, one major energy company elected to start small and build a limited capability before scaling out across business units and geographies. Its starting point was a selection of sites for an initial pilot. Each site piloted 3–5 categories of spend, with 10–15 suppliers and about 100 users. Third-party content providers were used to convert complex or high volume content. The result was the demonstration of a live tool and lessons learned that were incorporated into the subsequent rollout to hundreds of sites and tens of thousands of users.

Another company in the same sector took a similar approach, enabling it to discover that the predicted benefit case for the scaling phase could only be achieved if it placed greater emphasis on coordinating the eRequisitioning rollout with cross-business unit sourcing decisions. These cross-business unit sourcing decisions resulted in shared supplier catalogues at the local, regional or global level based on business requirements and supplier capability.

eRequisitioning providers
Hundreds of companies offer eRequisitioning software, content conversion and content aggregation for horizontal (cross industry) procurement marketplaces. Solution providers have entered this segment from a range of starting points:

- Market opportunists (e.g. Ariba and Commerce One)
- Enterprise Resource Planning vendors (e.g. SAP and Oracle, JDEdwards)
- Maintenance Management Systems (e.g. Maximo, BPCS)
- EDI vendors (e.g. Sterling Commerce, Harbinger)
- Component supplier management or data mining vendors (e.g. Aspect Development (now owned by i2 Technologies), Requisite Technologies)

Ariba and Commerce One's original aim was to dominate the eProcurement market by providing best-of-breed eRequisitioning solutions. The former initially focused on buyer-side catalogues and a world-class user interface, while the latter developed a strong conceptual vision of intermediation between buyers and suppliers through its focus on eMarketplaces. The two companies' distinct models were accompanied by different approaches to pricing – Ariba's pricing model is oriented towards an up-front buyer investment; Commerce One promotes a model in which buyers and suppliers pay on a transaction basis. Both companies have experienced enormous growth, albeit with fluctuating stock prices, ready access to capital and aggressive partnering. And both have acquired applications and talent to develop a broader solution footprint and formed alliances with consultancies to accelerate solution development and implementation. Their success in the eRequisitioning market, coupled with wins in major B2B eMarketplaces, have made these two companies market leaders.

Metiom, Inc. formerly Intelisys Electronic Commerce, Inc., has also emerged as a leader in the arena of building many-to-many eMarketplaces for global 2000 companies with its open B2B eCommerce solution. Metiom ConnectTrade enables large corporations to establish online exchanges and by using Metiom's patented Supplier-Managed Content and 'Wizard' approach to building electronic catalogues, small and mid-sized supplier bases are quickly able to participate in eMarketplaces as both buyer and seller. Metiom has enjoyed significant growth in the many-to-many eMarketplace arena and is backed by strong partnerships with technology leaders such as Hewlett-Packard, IBM, and Microsoft, systems integrators including Accenture and KPMG and major suppliers such as Staples and Office Depot.

Oracle and SAP have built eRequisitioning tools as extensions to their leading Enterprise Resource Planning (ERP) platforms. Oracle was quick to develop a comprehensive vision and leverage its resources to develop an Internet procurement capability (iProcurement) with rich functionality based on its ERP heritage that includes spend analysis and sourcing support in addition to strong eRequisitioning functionality. Oracle has also scored some important B2B eMarketplace wins and has made its main rival SAP look sluggish in the eProcurement marketplace. SAP has recently linked up its SAP Markets offer with Commerce One. By leveraging their extensive ERP installations, Oracle and SAP enjoy a pricing advantage within their existing customer base.

The effort required to create and maintain data quality and consistency in electronic catalogues for eRequisitioning solutions has often been underestimated. Third parties are often the lowest cost route to developing complex or high volume electronic content. Companies such as Aspect Development (now part of i2), Harbinger and Requisite Technologies have developed content rationalization capabilities to cleanse suppliers' item level data and organize it a meaningful way. Each company has taken a different approach to providing volume conversion and domain expertise (expert input to rationalize information into technically robust and user-friendly formats). In Aspect's case, low-cost labour in India is used to convert paper formats into electronic content. Requisite digitally re-masters supplier catalogues to present text and drawings as conceived in the original paper-based format. To meet the enormous complexity required to convert a user's search terms into a close match, search engines such as Requisite Technologies' Bug's Eye have been employed. Other such tools have been developed by companies such as OnDisplay and iMerge to enable real-time cross-catalogue searches where terms in individual catalogues may vary but match the user's request. This helps to reduce the rationalization effort required to convert all data into a common format.

In the future, the challenge of content management will be to enable interoperability between sources of content, increasingly driving the need for suppliers to provide electronic catalogues.

Adoption of eRequisitioning solutions
Penetration of eRequisitioning solutions is far lower than the hype and press announcements might indicate. As of September 2000, less than an estimated 1 per cent of Global 2000 companies' eRequisitioning spend was online. There are a number of factors contributing to this sluggish penetration. While US companies have been aggressive adopters, companies elsewhere have been focused on understanding and evaluating eProcurement solutions, and are not yet implementing the capability. Many companies also desire a single system and the current solutions cannot cover all spend, for example direct spend. The emergence of B2B eMarketplaces has also shifted some of the focus away from internal eRequisitioning initiatives.

The link between eRequisitioning and strategic sourcing, which should be a strong driver, has not been explicit to date and some early adopters have built a strong technical capability without achieving the full benefit potential of putting the best sourcing arrangements online. However, early experiences remain positive and benefits have been achieved through a reduction in purchasing administration time – time that can be redirected to strategic sourcing efforts.

For suppliers, the emergence of eRequisitioning has been viewed as both an opportunity and a threat. Companies have the opportunity to lock in and extend sales by embedding their products and services into customers' tools and by their ability to take accurate orders online. However, the cost of developing content or subscribing to eMarketplaces can offset this opportunity, especially when the company must still publish costly paper catalogues until all its customers are online. Suppliers fear the challenge of publishing the same content in multiple channels for different users. The availability of comparison shopping software has the potential to introduce a plug-and-play mentality to commodity purchasing.

There is likely to be a rapid expansion in the deployment of eRequisitioning solutions, primarily driven by the need for companies to manage transaction volumes through the B2B eMarketplaces that they will help to create.

Alternative business models for eRequisitioning
eRequisitioning solutions are not appropriate for all organizations. Some will not be willing to pay the price of a standalone implementation. These companies will look to other enterprises to provide alternative solutions that more closely match their business requirements and investment ability. There are two leading types of business model for this segment of the market. Banking-led buying groups for small and medium-sized enterprises (SMEs) represent a growing segment in the eRequisitioning market. Smaller organizations have little control or leverage over the indirect spend area, so the value proposition for eRequisitioning is only compelling if application and content development costs can be justified. Hosted services for SMEs will grow as Application Service Providers (ASPs) offer pay-per-use services that deliver all of the benefits of eRequisitioning without the up-front investment cost. Large companies with existing SME customer relationships will partner with other service providers to deliver value. Banks such as Barclays in the United Kingdom have entered this space with an eRequisitioning service that is part of an overall eCommerce offering to mid-range business customers.

Similarly, horizontal procurement eMarketplaces will combine the indirect spend of many large organizations through a shared eRequisitioning solution. Horizontal service models like those offered by Alliente and ICG Commerce (which now includes ePValue) provide hosted services that aggregate their clients' demand to negotiate improved supplier price agreements that are extended back to its clients. They have the opportunity, with specialist procurement resources, to build a wealth of procurement intelligence across a wide array of commodity groups.

eContracting solutions

eContracting is the ability to identify sources of supply online and contract with them directly over the Web to significantly reduce sourcing cycle times. eAuctioning currently is the most prevalent and popular of these solutions.

Online auctions are commonplace in the B2C arena; sites such as eBay and Priceline.com have developed the ability to offer consumers real-time price negotiation. This technique has even greater potential in the B2B arena, especially for companies looking to sell excess inventory or raw materials for short-term gains. In the early stages of its eCommerce strategy, one company auctioned requirements for food ingredients online and achieved a 35 per cent reduction in price paid (see Figure 3.2). Buyers may expect such benefits to cost reduction in addition to opportunities for supplier consolidation, access to improved sources of supply and increased efficiency of RFI/RFP/RFQ and bid processes. Suppliers may expect to benefit from introductions to new business opportunities, fair competition in the bidding process, increased market knowledge and ease of responding to RFQs.

eContracting providers
The solution landscape for eAuctioning is currently segmented into providers of standalone eAuction software, such as Moai and

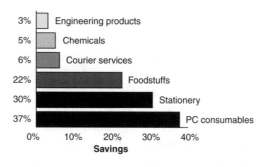

Figure 3.2 Percentage savings vs. historic price from dynamic pricing solutions.

DigitalMarkets and providers of eAuction services such as FreeMarkets, that outsource the preparation and execution of the auction process. Additionally, leading eRequisition software providers have acquired eAuction technology companies to complement their eRequisitioning offerings. For example, Ariba acquired TRADEX Technologies and Commerce One acquired CommerceBid.

Moai's solution, LiveExchange, provides the technology to conduct reverse auctions and online negotiations. It allows buyers and sellers to come together in real time to negotiate transactions. Bidding, negotiation and clearing rules can be customized to meet the specific business requirements of customers.

Freemarkets conducts online auctions for industrial parts, raw materials, commodities and services. In these auctions, suppliers compete in real time for the purchase orders of large buying organizations by lowering their prices until the auction is closed. The success of an online auction, which may take as little as 45 minutes to complete, is dependent on weeks of preparation. The planning includes: careful specification of the goods and services within each lot, specification of bidding procedures (visibility of bids to participants, sequence of bids, allocation of lots to bidders), pre-qualification of bidders and training and back-up procedures to ensure that all bidders are able to participate effectively.

Service providers operating in this space stress the importance of supply market knowledge and capabilities in sourcing strategy development. Such knowledge and skills enable the providers to identify and pre-qualify new bidders, as well as translate a lowest, total-cost sourcing approach into clear terms for an online event. These services are a critical differentiator for end-to-end service providers, as the full-value potential of online auctions can only be captured when there is sufficient competition between bidders and the event is conducted in line with best practice strategic sourcing principles. In the next 2–3 years the end-to-end service model pioneered by Freemarkets will become more prevalent – particularly in B2B exchanges. These sophisticated operators will provide expertise in specific vertical markets and complex auction types (e.g. multi-stage, multi-dimension auctions where price is one of many factors included in a series of related online events) as the auction tools themselves become commoditized.

Additionally, service providers like B2eMarkets integrate the benefits of an eEnabled procurement process with the eIntelligence made possible by the Internet. This type of offering has created a new position in the solution landscape – a hosted procurement service that can be integrated with a client's eProcurement system to provide tools to manage the sourcing process and combine internal management information with externally available intelligence. The benefit of this model will be the standardization of the strategic sourcing process across enterprises combined with the use of internal and external data to inform the supplier and product selection. As a Web-based offering available through an intranet, the solution will enable the collection of data about the efficiency and effectiveness of the overall sourcing and procurement process in addition to the measurement of transaction times and volumes currently provided by eProcurement solutions.

Adoption and reaction to eContracting solutions
Just as eProcurement kick-starts companies' B2B eCommerce programmes, eAuctioning kick-starts companies' eProcurement programmes. Adoption of eAuctioning is accelerating quickly as procurement departments use software providers and B2B eMarketplace hosted auction capabilities and associated sourcing expertise to deliver early successes – for example, one energy company saved over 60 per cent on listed prices from the incumbent supplier for its electrical goods. This can be a compelling low-cost, high-value proposition for buyers. While successful collaborative buying is notoriously difficult and early reactions from industry regulators suggest that eAuctioning will be limited in exchanges, this should not stop companies using this technique independently of each other.

The introduction of eAuctions has prompted a mixed response. Suppliers view the process as an opportunity for new business as well as a threat to the supplier partnership model. In the short term, eAuctioning shifts the balance of power to buyers. Early use of eAuctioning has had an impact on many relationships between companies, generally not for the better. Many suppliers have been shocked by the way in which the value of long-term relationships has been ignored by early eAuctions. For some there has been the realization that those relationships were not actually all that important. Others have reacted and tried to intervene to maintain the relationship with their customers. In one instance, a supplier offered significant savings on the renewal of a contract that had previously been eAuctioned if the buyer agreed not to take the renewal online.

eIntelligence solutions

There has been an enormous explosion in the amount of information delivered by Web-based technologies – but this is not intelligence. Intelligence requires the combination of sources of data and quality of data to provide as complete a picture of the dynamics of a business environment as possible. The reality is that much of the current available data is immature, unfocused and not sufficiently informative.

eIntelligence refers to the identification, collection and use of internal and external data to enable procurement to make smart sourcing decisions. It requires the combination of data mining as a one-off or internally developed capability with news-feeds – dumb through to intelligent – that balance internal and external intelligence.

Most organizations deploying eProcurement solutions opt for a reporting package that will extract data from both eProcurement and ERP systems to demonstrate workflow and audit trail data, enabling higher levels of visibility to identify cause and effect issues. This allows reporting flexibility to determine key performance indicators (KPIs) and the ability to monitor progress towards achieving metric goals, whether in transaction processing volumes or target savings levels per buyer.

eIntelligence providers

eIntelligence is a rapidly growing market in its own right. Already niche information providers are emerging to report on specific events where real-time information has explicit value. Web sites such as Water Online provide a portal that brings together UK water industry professionals, and includes information on procurement. The Thomas Register of American Manufacturers provides a similar function across several industries. The future holds further innovation in this area, which will bring many benefits to the procurement organization. Already the technology exists to create automated Web crawlers that will surf the Web in search of predefined information, such as the latest suppliers of a particular part in a particular location. However, even though the technology is willing, the standards remain weak, and without the widespread agreement and adoptions of standards which define parts, suppliers etc., it is unlikely that these tools will reach their full potential. The growth of private networks has allowed the reuse of procurement information generated within the network for the benefit of other participants in the network. Companies can gain clear advantages by sharing insights on marketplace and price developments.

Companies should think about their commodity, performance and marketplace data requirements in order to develop an eIntelligence strategy that will underpin the eProcurement process and add to the skills and capabilities-building that is required to support the solution.

Emerging eIntelligence solutions are important to the development of a value-driven procurement capability because they extend eProcurement across the steps involved in identifying opportunities and placing contracts. These new tool-sets are essential building blocks for delivering value through procurement. Organizational development is a key to ensure that the tools are fully exploited by individuals with the right skills and knowledge to develop sound sourcing strategies and innovative approaches to supplier and supply chain integration.

Adoption of eIntelligence solutions
It is still early days in the adoption of eIntelligence solutions for procurement. Most companies have yet to exploit the wealth of information that can be generated by a variety of new systems and access to Web-based exchanges and knowledge resources. The introduction of checkout scanners to supermarkets is a good analogy. The first priority was to get the scanners working and the employees trained. Analysis of the data collected followed. With eIntelligence, the first priority is to get the systems working; knowledge and insight will follow.

Demonstrated value and critical drivers for successful implementation

eProcurement and strategic sourcing are linked in a potential virtuous circle of reduced cost and improved performance based on the linkage between four value levers (see Figure 3.3):

● Reduce administrative costs
● Enable better sourcing and supplier management
● Rollout and sustain sourcing deals
● Enable (and share) supplier benefits

With enough management attention and effort within the procurement organization, and enough goodwill, each of these goals could be achieved without the development of eProcurement capabilities. However, the reality is that procurement generally is not a high priority for corporate

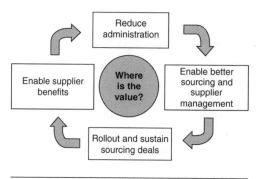

Figure 3.3 Four value levers of
eProcurement and strategic sourcing.

users and the burden of administration has prevented many
organizations from effectively implementing strategic sourcing
programmes.

Reduced administrative costs

Time spent on requisition-to-pay activities is estimated at over $100 per
transaction. The amount of low-value processing (e.g. manually
converting requisitions into orders or addressing the paper chase
currently associated with invoice processing) is in excess of the value of
many orders. eRequisitioning enables corporate users to buy efficiently
(online order placement may be slower than phone-based ordering, but
there is less up-front checking to do), remove non-value adding steps in
the procurement process and eliminate the usual order and invoice error
processing. eProcurement tools may be used as a focus for process
improvement – for example, rollout of self-billing and evaluated receipt
settlement can be enabled rather than delivered by the new tools.

The cost of an eRequisition is therefore comparatively low – estimates
fall in the $10–$30 range. In addition, circumventing the invoice paper
chase means that suppliers can be paid according to contract terms.
With legislation moving towards penalties for late payment, the ability
for buyers to pay on time will become increasingly important and
valuable.

A comprehensive eProcurement solution can also reduce the significant
administrative effort associated with making sourcing decisions, such as
investigating spend category research, time-consuming paper-based

RFQs and RFPs with suppliers and protracted negotiation cycles detract from strategy formation and decision-making. eIntelligence reduces the amount of time required to gather and analyse information and eContracting automates low value-added tasks such as RFP evaluation to allow the procurement professional to review ranked electronic responses rather than sift through and re-key supplier information.

The productivity savings from reduced administration do not all hit the bottom line – but improved productivity enables procurement professionals to focus more time on better sourcing and supplier management.

Improved sourcing and supplier management

In many instances spend category analysis is cursory. Generally, an estimate of historical spend without any forecasting or segmentation of demand is used as a basis for contracting with suppliers, resulting in poor sourcing decisions. Additionally, poor market information reduces the opportunity for buying companies to understand their possible leverage points in negotiation. Finally, once a sub-optimal deal is in place, little or no fact-based performance information is available.

eRequisitioning plays an important role in improving sourcing performance. In one instance, a sourcing deal for journal subscriptions within a large organization was based on high level estimates of spend of around £0.5m. This figure was derived through a painstaking process of trawling through account codes for trade publication and book purchase entries. The deal struck was based on this overall volume of spend, but the organization did not know the publication titles and many staff members had subscribed separately. The spend was too fragmented to perform research by user in order to identify duplication, so eRequisitioning was used to channel subscriptions through a service provider. Details of individual subscriptions were captured as each order was placed and eIntelligence analysis after 6 months of operation revealed that 20 per cent of the spend was on duplicate publications. This saving could only be achieved through the visibility provided by the eRequisitioning tool.

eContracting reduces sourcing administration while also delivering incremental benefits. In some situations, online reverse auctions enable greater supplier participation and increased visibility of market prices.

Where there is sufficient competition within the supplier base, real-time feedback enables the most competitive bidder to put forward a winning offer. There are fears that in some instances, especially in a non-competitive market, the winning bid may not represent the supplier's best price (indeed the lowest price may not win the business). While this is a possibility, observation has shown that when there is a competitive supply market, results from online bidding exceed historical results and expectations by 10–20 per cent, suggesting that the tool enables additional benefit delivery.

eIntelligence about suppliers' performance can be used to identify further improvement opportunities within the customer or supplier organization. Delivery performance for inventory items is relatively easy to compile from ERP systems; however, delivery of services is more subjective and dependent upon user feedback. eRequisitioning can be used to gather such performance information. The desktop receiving process includes simple pop-up feedback boxes, which poll users once goods or services are due or delivered. These benefits are opportunistic and therefore difficult to quantify.

Rollout and sustain sourcing deals

Benefits that have been negotiated through better sourcing contracts typically phase gradually. Often, initial communication to users about a new deal does not succeed, purchasing to contract diminishes over time and the supplier base proliferates.

eProcurement accelerates the rollout of new deals by embedding the new arrangements in electronic catalogues or providing easy reference contracts databases for those items or services that cannot be eRequisitioned. Terms become effective as soon as a new catalogue is in place, ensuring that the benefits are achieved immediately. Providing users with an easy-to-use tool also reduces the proliferation of the supplier base, as users do not have to look up correct supplier and item details. For areas like training, which are highly fragmented in many organizations, the benefits of an initial supplier rationalization exercise are typically lost over 3–5 years as the organization begins to spend with tens (or in some cases even hundreds) of suppliers. The benefit of eProcurement in this instance increases over time as strategic sourcing benefits that would have diminished are retained.

Enable supplier benefits

Reduced administration in order-taking through accurate online orders for eRequisitioning customers, sustained sales through increased compliance in the buying organization, improved visibility of market prices through online bidding and fact-based performance feedback from customers all improve a supplier's ability to meet customers' needs at lower total cost.

Most procurement departments aspire to have more strategic relationships with fewer suppliers. While this does not lead to a partnership in every case, it does require the procurement department to commit to professional and efficient daily operational contact with suppliers. eProcurement supports this goal with the ability to order accurately, deliver expected spend volumes and pay on time, thereby delivering financial benefits through shared savings. eProcurement enables procurement departments to focus more time on value-added activities and equips procurement professionals with the tools to deliver the best commercial arrangements in an effective manner (see Table 3.1).

Trends in eProcurement and strategic sourcing

Three main trends will act as fundamental drivers in the realization of value from eProcurement:

- eProcurement tools will shift from transaction processing to decision support.
- eMarketplace developments will drive multi-level sourcing decisions.
- Suppliers who understand their cost-to-serve will dominate.

Shifting from transaction to decision support

eRequisitioning has enabled large numbers of users to conduct self-service purchasing from corporate contracts. Companies that have always achieved procurement excellence and applied the best practices of strategic sourcing to their supply bases are best positioned to take advantage of eProcurement and the investment in eRequisitioning. The

Table 3.1 The added value of eProcurement.

Where's the value?	How do I get the value?		
	eRequisitioning	eContracting	eIntelligence
Reduce administration	● Improve productivity by eliminating paper-based processing. ● Allow procurement department to focus on value-added activities.	● Improve productivity by eliminating paper-based processing. ● Enable simultaneous multi-supplier negotiations.	● Reduce time spent finding internal and external data.
Enable better sourcing and supplier management	● Provide detailed spend data (item and user level) for fragmented categories. ● Provide performance information through desktop receiving.	● Deliver an up-front step-change benefit, followed by incremental benefits through online bidding. ● Enable focus on value-added activities through automation and decision support. ● Reduce sourcing cycle time and get more of the supply base strategically managed from the same number of procurement professionals.	● Improve quality of internal and external information to leverage in negotiations. ● Provide performance information to drive supplier development.
Rollout and sustain sourcing deals	● Enable instant take-up of new contracts by making new terms available to entire organization in the electronic catalogue. ● Prevent supplier proliferation by embedding preferred suppliers in system.	● Increase take-up of spend outside of eRequisitioning through contract databases.	● Report compliance to enable corrective action to be taken.
Enable supplier benefits	● Improve productivity by eliminating paper-based processing. ● Reduce errors. ● Guarantee spend volumes.	● Improve productivity by eliminating paper-based processing. ● Improve visibility of market information through online bidding.	● Provide performance information to drive supplier development.

added information provided from eRequisitioning is simply too costly to capture without eProcurement, and takes time to become truly useful in decision support of the procurement process.

The bigger prize, for organizations with an industrial bias, lies in the exploitation of eIntelligence and eContracting to deliver significantly greater value from direct expenditure. These solutions will deliver spend history, demand forecasts, commercially significant supply market information, and will support world-class sourcing techniques with automated tendering and dynamic pricing models. The effect will be to provide an integrated suite of decision support and automated sourcing tools.

The array of eIntelligence and eContracting tools in the marketplace today will consolidate into two or three comprehensive solutions. These solutions will provide dynamic, configurable, sourcing support systems that span opportunity identification, contract placement and supplier/ contract management.

The difference between the current focus on narrow solutions and an advanced vision for eProcurement is highlighted through several examples:

- **Auto-intelligent supply review**: intelligent agents search the Web for events that trigger the review of a supply relationship. These events – new or lower cost alternatives to current arrangements – discovered by the agents trigger the start of a strategic sourcing cycle for the buying organization. A suggested project plan including a summary of spend history, a forecast by service or item and key supplier information are suggested to the buyer who can then make a 'go' or 'no-go' decision.
- **Pre-configured supply options**: for more dynamic or time-sensitive markets a hybrid between a full supply review and real-time exchange will emerge. In this instance, buying companies will be able to place options to purchase goods and services on suppliers' Web sites on the basis of price and other variables, and automatically trigger bulk purchase of an optimized quantity or a change of supplier when its conditions are met (also called definite auctions).
- **Intelligent infomercials**: the current generic level of industry news- feeds targeted to desktops will evolve into specific supplier infomercials – product development news including industry or company-relevant performance statistics will be delivered to specific buyers within target customer organizations. In these infomercials, three-dimensional representation of products in operation will be complemented by the ability to model buyer operating conditions and review results for a range of operating scenarios. This will impact both the timing and cycle-time of the purchase decision as well as the buyer-side time-scale for test and release into full production.

- **Multi-stage auctions**: buyers will be able to identify and target distinct value-added areas in the sellers' total value proposition at a more granular level. This in turn will lead to successive rounds of evaluation within the same auction event. For example, in procuring professional services, a buyer may request bids for marketing services based upon:
  - the composition of the account team and consistency of the team
  - their daily or hourly rates or the ability to work at a fixed cost
  - their willingness to accept penalties for delays
  - leverage of media buying power
  - a quantitative track record of improved penetration or sales

Furthermore, these opportunities to pre-configure supply arrangements or exploit dynamic pricing models will be extended across members of buying consortia. Within these consortia the cross-enterprise configuration will be defined on a commodity-by-commodity basis where the buyers have similar or complementary supply requirements. All these effectively encourage the supply base to be more efficient in their operations and interactions with their customers and suppliers. Hence, expect to see these initiatives spread up and down the supplying tiers.

Multi-level sourcing decisions

As B2B eMarketplaces mature they will form three types of relationship with related trading communities: compete, consolidate or collaborate. Competition with, and consolidation of, eMarketplaces is directly comparable to traditional corporate strategy – eMarketplaces will measure success in terms of profitability and market share and waves of merger and acquisition activity will take place. The potential for collaboration between eMarketplaces is dependent upon open standards across the technologies used by each community. An increase in open standards will enable horizontal scaling, which in turn will reduce switching costs and lead to a focus on delivery of value through the value chain. Successful B2B eMarketplaces will need to create 'stickiness' by delivering superior performance across the entire value chain and identifying and exploiting key leverage points into related networks.

Domination of cost-to-serve suppliers

eProcurement combined with strategic sourcing will lead to visibility of total supply chain costs. The network effect will drive out total network chain cost opportunities in the short-term, and total network cost visibility in the longer-term. As companies begin to understand their costs better, we expect to see new forms of 'co-opetition' between suppliers within networks in which small numbers of competitors collaborate to fulfil the demand of one company.

The improved ability to search and select new suppliers through the Web will lead to new supplier selection criteria including support for new transaction types and value-added offerings. Membership in eMarketplaces already is dependent in part upon the eCommerce capabilities of potential participants. This will evolve quickly into a supplier's ability to support value-added services such as eFulfillment and ePayment, and has potentially far-reaching consequences for those companies who are followers into B2B eMarketplaces.

Suppliers will risk elements of current competitive advantage to develop shared, network-based barriers to entry within an eMarketplace. The driver for this collaboration will be the opportunity to scale these new offerings horizontally into related networks or chains, exploiting new areas of advantage and creating barriers to entry. This development will underpin the interconnectivity of multiple eMarketplaces and the development of global alliances between trading-partner eMarketplaces. Early indications of this trend are seen in the early adoption of online logistics collaboration; Internet-enabled collaborative planning, forecasting and replenishment and the development of auctions for time-period-based demand.

We expect that those suppliers who understand their cost-to-serve will dominate. Their closer understanding of the true value that they can add will aid the decisions they make on critical business issues. They will, at the very least, know when they are pricing for profit or for market share gain.

Conclusion

eProcurement is not an end in itself. Rather, it is a foundation on which companies are building much broader eCommerce capabilities – an early stepping-stone on a journey to create eEnabled supply chain capabilities. Many companies have approached eProcurement as a low-cost, low-risk initiative, but very few have fully understood the scope and value of what they are trying to achieve. Most have found eProcurement has involved higher cost and risk than anticipated. Nevertheless, the market leaders have been effective in turning eProcurement into a bridgehead for shaping the future supply chain, building cross-company collaboration capabilities, and achieving B2B eCommerce.

The hype and media attention devoted to the arrival and development of eProcurement and eMarketplace activity is above all, an excellent example of marketing and advertising money well spent. The early eProcurement pioneers – Ariba, Netscape and Commerce One – had no installed customer base to leverage and nothing to lose other than a great opportunity. They recognized the weakness of existing ERP providers and moved quickly into this space. The rapidly evolving 'land-grab' – of both customers and operating space for Web-based procurement and marketplaces – has seen these nimble companies carve out a sizeable market niche.

The arrival of eMarketplaces as the next wave of supply chain collaboration has raised questions about the scalability of these initial offerings and accelerated the development of strategic alliances. It has led to the development of alliances between old and new players to sustain market presence and leverage complementary products and services – for example, the alliance between IBM, Ariba and i2 Technologies and that between SAP and Commerce One. Against this background we expect the rate of adoption of eProcurement solutions to accelerate over the next two years as solutions mature and their application to commodities is better understood.

The procurement of indirect materials has long been an area characterized by uncontrolled spending, insufficient accountability for decision making and multiple suppliers with a wide range of commodities. eProcurement has made the most impact in this area, but the principles have been slow to migrate to the procurement of direct materials. However, in the longer term, we expect to see the increasing application of these tools to direct spend categories as eProcurement and eMarketplace solutions become more mature and as companies improve their ability to exploit their value.

eProcurement and the emergence of eMarketplaces will have far-reaching implications on the roles of procurement professionals, minimizing time spent on administrative tasks and requiring them to acquire new skills. Tomorrow's procurement professionals will broaden their focus to encompass the extended supply chain and management of the configuration of eProcurement tools, including business rules and electronic content.

4 The eFulfillment challenge – the Holy Grail of B2C and B2B eCommerce

Introduction

The major prize-winners in the eEconomy will be those organizations that overcome the challenge of fulfilling eCommerce-based demand. If the model is right, becoming the contract logistics provider of choice for electronic traders that outsource their fulfillment capability could be a winning move. Providing such services has the potential for tremendous success, particularly for companies with extensive fulfillment expertise.

We use eFulfillment to describe the unique aspects of delivering products and services to eCommerce customers. It incorporates activities such as order management, call centre management, online credit checking and credit card processing, inventory management, warehousing and shipping, and processing and disposition of returns. Together, these activities represent a significant proportion of eCommerce companies' tangible assets and have the greatest influence on customer service.

In this chapter, we examine the changing role of fulfillment in the eEconomy. The ability to satisfy consumer demand effectively has challenged online retailers for some time now, and with the emergence of eMarketplaces, eProcurement and other B2B Channels, eFulfillment is becoming a critical enabler of B2B eCommerce. Companies are progressing toward an integrated B2B2C eFulfillment approach with a broader end-to-end supply chain performance objective. The key questions we address in this chapter are:

- What is the role of eFulfillment in the eEconomy?
- What are the new B2C eFulfillment models and solutions?
- What is new in B2B eFulfillment models and solutions?
- What new eFulfillment business models are emerging?

The role of eFulfillment in the eEconomy

For many companies engaged in eCommerce, profitability has been compromised by an inability to effectively manage and fulfil customer orders. The physical delivery of products to customers is one of the most significant challenges for B2C and B2B businesses alike, and fulfillment capability is already proving to be a strong market differentiator. Good reliable service is a key factor to achieving repeat sales while poor service creates loss of trust and corrodes future selling opportunities. The customer service paradigm in the value chain network, consequently, is changing. In the future, the basis for measuring eFulfillment service will be percentage of perfect fulfillment executions rather than average performance.

Additionally, value-added services are fast becoming entry ticket requirements that are assumed to be standard offerings. These include:

- real-time inventory visibility and product availability, including production capacity
- real-time visibility of accurate order status with package tracking capabilities
- real-time online payment approval
- easily accessible, responsive customer service and support before, during and after the sale
- small package shipments and trailer loads
- single consolidated shipments
- reliable, affordable delivery with different drop-off options – dock, store, home etc.
- hassle-free product return processing and payment refunding

Achieving these capabilities requires integrating Web-based customer interfaces with traditional inventory, order processing, warehouse management and transport management systems. The complexity of knitting together such systems to provide accurate real-time data is stretching current technology to its limits.

Leading organizations recognize the importance of logistics and the systems application footprint, and view supply chain and technology as strategic board-level issues requiring executive-level understanding. In developing an eCommerce strategy, as much consideration should be given to the fulfillment strategy as to the choice of product offering, marketing approach or Web site design.

B2C eFulfillment

When the 'transact' stage of Internet business began, the virtual component of eCommerce attracted the majority of attention: the ability to generate Web site traffic and turn it into paid orders. Many online retailers, particularly those that had well-established distribution capabilities, were crushed by the mistaken assumption that eCommerce was just another way to take orders, and that the fulfillment of these orders could be handled by the existing infrastructure, information systems, processes and skills.

One of the biggest impacts the Internet has had on retailing is to transform products that are traditionally purchased in physical stores into seemingly virtual products that can be obtained on demand, at the point of use, and even be priced on a per-use basis. Despite this focus on the virtual world, a huge portion of Internet commerce still revolves around presenting, selling and delivering physical products to consumers and businesses. eFulfillment is the essential link between the virtual and physical worlds. It has also proven to be the Achilles' heel of many businesses using the Internet to reach their customers.

Customers expect a level of performance, speed, and precision that is significantly beyond their performance expectations for more traditional businesses. Compounding the effect of high expectations, Internet technologies make it possible for customers to access vast amounts of information, and customers expect ready access to the status of their orders. Every step, or misstep, is visible to the customer, often as the process is occurring. This visibility places a tremendous premium on online retailers being both reliable and fast during each phase of their operations.

In the world of the Web, everything is dynamic. This is potentially true of the products and services offered to customers. It is quite literally possible for an online retailer to change its entire offering in a matter of minutes, which from a marketing perspective is a distinct advantage. However, this advantage can wreak havoc on carefully designed fulfillment processes, material handling systems and warehouse buildings. These physical processes suffer from the limitations of physical matter, which, unlike a Web site, cannot change instantly to accommodate a new offering.

Consumer-direct models put high demands on the distribution system as a result of smaller, more frequent, more time-sensitive orders to an often significantly broadened geographic base. The pressure is further exacerbated by product returns processes. Complete change in logistics operations needs to occur for both a manufacturer with no one-to-one shipment capability and a regular retailer with traditional outlets.

B2C eFulfillment models

eFulfillment solutions must take into account a multitude of requirements that can be combined into an array of combinations. Subtle but significant differences exist in the way operations must be carried out even for companies with good capability and experience in traditional one-to-one shipping. We have identified eight B2C models, supported by two call centre options (see Figure 4.1) determined by ownership of product, physical assets and the point of delivery to the consumer.

Call centre options
A. Insourced

Insourced call centres are ideal for those companies that operate in a complex selling or customer service environment, or that have sufficient business volume to absorb the high fixed costs. The benefits include control of the process, quality and customer relationship; a high degree of service and process flexibility;

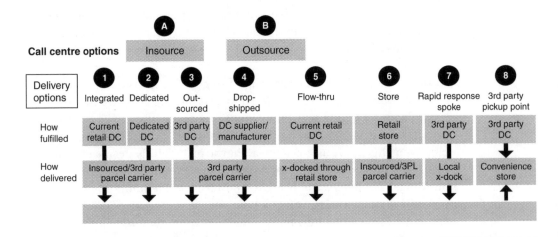

Figure 4.1 B2C eFulfillment models.

optimum use of technologies to lower costs and, potentially, the lowest operating costs.

B. *Outsourced*

Outsourced call centres are ideal for companies with no existing call centre asset, low volume, or a relatively simple selling and customer service situation. Key benefits include rapid start-up, and low initial and ongoing costs that are proportional to volume.

Delivery options

1. *Integrated*

Integrated fulfillment involves building eFulfillment capability into an existing bulk distribution centre. Though tempting, this tends to be problematic for many retailers due to issues of channel conflict – specifically, the difficulty aligning pick and pack processes for Internet orders with retail store channel consignments.

2. *Dedicated*

A dedicated fulfillment centre involves building or acquiring a fulfillment capability that is dedicated to eCommerce trade. The primary barrier to this model for most traditional retail organizations is cost. Amazon.com has aggressively shifted to this model.

3. *Outsourced*

Third-party outsourced fulfillment models often are faster to set up and are therefore attractive to start-up online retailers. Most small package carriers such as UPS and FedEx, offer some level of eFulfillment service capability. This ready access to proven technologies and reliable performance make this an attractive option; however, faith in the capabilities of the third-party provider and alignment of eCommerce strategies are key to success in this model.

4. *Drop-shipped*

Drop-shipped fulfillment, long favoured by cataloguers, may turn out to be one of the most cost-effective models. If the distributor has proven capability for one-to-one shipping, this model has many of the advantages of a third-party model. However, combining sale of product with fulfillment service at the distributor substantially shifts negotiation power away from the eCommerce company. If the value-chain relationships are managed equitably and the technology to synchronize deliveries from many points is improved, this model may predominate.

5. *Flow-thru*

   Flow-thru fulfillment, is currently used by Sears for online sales of big-ticket items. Internet customer orders are picked as individual orders at the distribution centre. They arrive at the retail store as individual customer orders and are cross-docked through the store to the customers. This model is effective for shipping products that require a substantial service component.

6. *Store*

   Store fulfillment, typified by Peapod's original grocery fulfillment method, has not generally been successful. Internet customer orders are picked from the retail store shelves. Technology integration across many fulfillment points and quality control are big barriers to the success of this model; however, it is attractive to existing retailers.

7. *Rapid response spoke*

   Rapid response spoke is an 'urban fetch' model designed to handle high-premium, time-critical delivery services – within the hour. The model is an instant gratification model. For example, a customer may order chips, nuts, soft drinks, wine and wine glasses for an impromptu gathering, or a video, ice-cream and some flu tablets for a sick evening at home. The offering incorporates a wide range of products, and inventory must be stocked near the customer base, so it is only viable in densely populated areas.

8. *Third-party pickup point*

   Third-party pickup points provide an alternative to home delivery models and offer consumers the flexibility of no delivery window at a lower cost. Customer orders are picked, packed and despatched to a local collection point, potentially by a third-party service provider. This model has the potential to offer simplified security and cash management as well as the provision of special handling requirements such as temperature control. Retailers such as Payless.com have explored this model – customers could pick up their online orders at any of the Payless shoe stores.

The major multiple supermarket arena in the UK provides several models.

CASE FILE

Tesco will spend £35 million in the year to July 2001, rolling out its home delivery and Internet service. The vans and picking and packing equipment needed to provide the service to 90 per cent of the UK population will account for the majority of this expenditure. Tesco aims for first-mover advantage to ensure that its online store provides everything consumers want from a home shopping experience. Both Tesco and Iceland have rolled out services based on picking Internet orders in stores rather than from dedicated warehouses to enable immediate national coverage. They retain the option to migrate to a dedicated channel or shared warehouse model.

Wal-Mart subsidiary, Asda, intends to roll out a national Internet service within the next five years. Asda believes that warehouse-based service is the only model that allows retailers to offer the level of service that customers expect.

A warehouse-based system has proved unsuccessful for Somerfield, which scrapped its trial £11m home-shopping service, known as 24–7. 24–7 had not grown at the expected rate and distracted management from Somerfield's core businesses.

CASE FILE

Based in Chicago, Peapod, Inc. is one of the USA's leading online grocery retailers with a national grocery delivery service. Replacing its in-store pick model with a fast-pick, regional distribution centre model has enabled Peapod to achieve significant efficiency gains.

Thousands of non-perishable grocery items and household goods are drop-shipped from Peapod warehouses. Delivery operations in over 20 fulfillment centres cover in excess of 8 per cent of US households. Peapod's national delivery programme has the potential to increase product revenues with little additional overhead by further utilizing existing market warehouses.

The company's database and customer profiles enable it to deliver highly-targeted, one-to-one advertising and promotions, and provide consumer goods companies with a forum for targeted interactive advertising, electronic coupons and extensive product research as well as substantial data regarding demographics and customers' purchasing intentions and behaviour.

In addition to produce, meats, bakery, deli and dairy items, Peapod offers a bundled goods programme called 'Peapod Packages' to deliver customized bundles of non-perishable grocery items for special occasions, including new baby packages and college cares packages. Peapod has several specialty shops that sell books, baby items, beer, wine, spirits, pet and seasonal items.

CASE FILE

Webvan Group, Inc. combines online grocery shopping with a personalized courier service that delivers products into customers' homes within a 30-minute window of their choosing. By the end of 2001, Webvan plans to open distribution centres in 15 geographic markets to serve a hub-and-spoke delivery system. Orders collected from the Webstore, are routed and managed by the distribution centre, transferred to stations and delivered to customers' homes. Webvan says this model enables it to efficiently and cost-effectively deliver consumer goods to the home by combining centralized order fulfillment with decentralized delivery. Webvan is investing millions in their specialized replenishment process that includes warehouses in cheap industrial locations, specialized picking systems and significant numbers of vehicles to optimize the way the product is delivered customers.

Webvan delivers consumer electronics and home entertainment products, clothing, books, groceries and household goods, and mass transit fares and toll cards. Webvan has developed new capabilities designed to enhance the shopping experience and its business model economics including the introduction of the 'Tell a Friend' referral programme, that rewards customers for introducing Webvan to their friends. The Web site was also enhanced with an expanded customer feedback feature that allows customers to make product suggestions at various locations on the Web site. In addition, broader 'add to cart' and 'add to shopping list' capabilities were added to the Sensations area of Webvan.com, which includes recipes and menus presented by leading chefs and nutritionists.

B2C eFulfillment solution providers

Getting all aspects of eFulfillment right is clearly important to successful Internet retailing. It will affect how likely customers are to buy from an organization and determine a substantial portion of the costs of the business. Given the varied eFulfillment requirements for different companies, several service providers, from national post organizations to third-party logistics providers to public parcel carriers and mail order houses, are jockeying for position. Some are applying traditional fulfillment services to the new requirements, while others are developing entirely new businesses for this segment.

Postal services

National post organizations, with a high drop density, have the clear advantage in B2C deliveries. In many areas, they currently are the only cost-effective way to serve a low- to mid-margin market. However, most traditional operating models can deliver only standard package sizes. So, if the first attempted delivery of a non-letterbox sized package is unsuccessful, the home delivery model becomes a pickup point model. Some of the more progressive national post organizations, such as Deutsche Post, TNT, Royal Mail, Parcel Force and Australia Post, have recognized the B2C opportunity and are developing new capabilities, such as after-hours and weekend deliveries, in an effort to capture the significant benefits.

CASE FILE

Following the success of late-night delivery slots, the UK's National Post Office is considering plans to offer same-day home shopping delivery in major cities. Royal Mail already provides an informal same-day service for Amazon.com in the UK. This is not an advertised service; rather, it is supplied as a bonus to customers in central London where several other same-day services such as UrbanFetch and Iforce make deliveries. National Post Office executives are using services such as these to position the group as a complete Internet fulfillment partner to compete against private distribution companies and overseas post offices. Most online shipments are serviced through its existing infrastructure in metropolitan areas. However, providing services to rural areas is a challenge due to higher costs involved – an issue that is fuelling debate over the 'digital divide' between people who can exploit the benefits of new technology and those who are denied access. Rural communities may have to be resourceful, providing their own solutions through collective drop-off points or leaving parcels with neighbours. However, with the take-up of Internet retailers being better than expected, a late-night, week-day parcel force may be extended geographically.

Third-party logistics and services providers

With expertise in transport and distribution, third-party logistics and services providers (3PLs) should be dominating the new world of eFulfillment. However, traditional logistics companies are generally lagging in developing capabilities to partner in dynamic supply chain networks. For many, managing contracts, assets and industrial relations have been core competencies, and they have less experience with new requirements for technology, planning and service development competencies. To remain competitive, these traditional providers will have to undergo significant organizational change. Many have created eCommerce divisions that are not integrated with the rest of the business, causing confusion about ownership of customers, varied service levels and minimal knowledge sharing, different entry points for customers and limitations in scalability. Despite current narrow profit margins, these companies will have to invest substantially to integrate and scale their eCommerce operations. Information technology is their biggest barrier to success – many lack the client/server generation of business systems that are common in most of their customers' organizations. Most 3PLs have single or occasionally dual user operations, use internal carriers and have limited geographic coverage.

Some of the major 3PLs are actively developing the technology and connectivity that enable them to offer a service to core customers and also to some of the marketplaces. Companies such as Exel Logistics have a digitally connected link in the Automotive industry in Spain and Mexico and Tibbet and Britten have a similar eEnabled service in the retailing industry.

Global parcel carriers

Global parcel carriers have all been investing considerable sums to integrate their business services to customers via the Web, spurred on by the attraction of the B2C market. TNT tracking can pinpoint shipment details from any Internet-based computer and the other majors have similar capabilities. UPS commercialized the use of bar-code readers and put computer tracking online before the B2C boom. This capability had started to move business from using the traditional postal services.

Technology development by the major global carriers used the Web capability to offer the reliable and openly trackable delivery performance required for long-distance package fulfillment for the B2C and B2B markets. FedEx's eCommerce Builder helps potential B2C customers build and manage an online store, including billing, payments and customer management.

DHL's new ThermoExpress service is an innovative solution to ship perishable goods. It offers a patented insulated container that provides 10 times the insulation of a typical styrofoam cooler plus the reliability and support of the DHL system. This service could not have been developed without the extensive use of the Internet to optimize the delivery timings.

Fulfillment costs are, however, a major obstacle for widespread use of these package logistics providers in the B2C arena. Some of the product distribution cannot support the full logistics cost, hence the growth of some local operators and the more extensive use of the postal services mentioned above. This need for a lower-cost fulfillment channel is being recognized and addressed by FedEx Ground.

CASE FILE

FedEx Ground is a $2 billion subsidiary of FedEx Corp. and the second largest small-package ground carrier in North America with an automated network of 369 distribution hubs and local pick-up-and-delivery terminals throughout the United States and Canada. More than 2.5 million customers are connected electronically through their information network and approximately two-thirds of its US domestic transactions are now handled online.

FedEx plans to achieve full US coverage by September 2002 when FedEx Ground completes the addition of approximately 150 home delivery terminals to increase the network's service coverage. FedEx Home Delivery was created to respond to business-to-consumer shipper demand with a better ground delivery solution for the residential market. To help online retailers address mounting fulfillment challenges, the new service offers an economical and customized residential solution to fit the individual needs of customers. The suite of service options includes extended evening delivery, Saturday delivery and premium services such as day-specific, signature and appointment delivery. FedEx Home Delivery also offers a money-back guarantee.

Mail order companies

Mail order companies have the potential to create an end-to-end offering that includes payments, call centres and physical fulfillment. However, they will have to improve their service levels to operate effectively in an eCommerce environment. Fulfillment houses are generally multiple carrier organizations. For the most part, they are relatively low in technology sophistication. Traditional catalogue retailers generally have high technology sophistication, but they face a market already saturated by their existing business and channel conflict with potential B2C customers. Despite their extensive network of home delivery, the catalogue retailers still need to enhance their fulfillment capabilities in order to meet the distinct demands of the Internet shopper. The retailers will need to achieve promised delivery time from a period of hours to a maximum of 2–3 days and to view the status of the delivery online.

The major catalogue organizations of GUS and Freemans in the UK and L.L. Bean in the USA are recognizing their position in fulfillment and strengthening their service offerings. The majors are acquiring some of the dotCom companies that are under cash pressure in order to bring a more entrepreneurial style to their core products.

eFulfillment specialists

Several specialist eFulfillment service providers are emerging and several traditional third-party firms have retooled themselves to concentrate largely, or exclusively on eFulfillment. Within the specialist ranks there are two categories of providers: those who focus on asset-based services and those who focus on information exchange.

Two examples of asset-based service providers are iFulfillment, in North America, and The Prolog Group, which serves the European market. Prolog offers a range of fulfillment-related services but have added Web-specific offerings to allow them to compete with fulfillment giants such as TNT and Deutsche Post, who increasingly offer a 'one stop shop' to eTailers. PFS Web is one of the few companies to operate eFulfillment facilities in both North America and Europe using shared IT infrastructure to provide systems resilience at reduced cost. NewRoads, also asset-based, evolved from traditional services to a service profile that match the needs of eCommerce through acquisition and implementing a comprehensive technology strategy. The substantial scale of their existing operations allows many of these companies to more easily absorb new clients. However, many firms are still reluctant to take on start-up clients because of business uncertainties.

Information-based eFulfillment providers include a variety of types of firm that can often look much like software companies. What distinguishes them is a reliance on technology to integrate complex networks of providers into a seamless fulfillment process. Therefore, their software offerings tend to be oriented toward interconnectivity rather than fulfillment functions. Some examples of these firms are electroneconomy, Yantra, Celarix and logistics.com. Typically companies in this class will be either warehouse or transport oriented. electroneconomy and Yantra, for instance, focus on warehouses, while Celarix and logistics.com concentrate on transportation processes.

As these information-based providers are more likely to be start-ups or very new ventures, care must be taken to ensure the reality of their proclaimed offerings, and that they have adequate levels of funding for continued operations.

CASE FILE

iFulfillment Inc. provides complete fulfillment services engineered exclusively for eCommerce transactions, drawing on sophisticated supply chain execution capabilities to integrate fulfillment and delivery services into online retailer capabilities. Web-based tools enable online retailers to deploy fulfillment services quickly and manage and monitor supply chain activities in real time.

Without standard fulfillment technologies or practices, online retailers have had to turn to traditional distribution companies without eCommerce savvy or make large investments to build a proprietary fulfillment solution. iFulfillment fills that void with a standard-setting solution that incorporates open standards such as XML and unites online stores with iFulfillment's operations and parcel carriers to provide highly efficient, seamless order fulfillment. eCommerce companies enjoy lower costs because iFulfillment aggregates their buying power for distribution space, and uses state-of-the-art materials handling technology, shipping services and packing materials.

A number of eFulfillment companies have specifically sought to address the issue of deliveries failing due to the customer not being at home when delivery is attempted. Two main models are being tested: the first is based on the collection point concept, and the second is based on secure unattended delivery containers.

The developers of the collection point concept such as Collectpoint are providing the integration layer between eTailers and the various convenience stores and fuel filling stations that they have signed up as collection points. While this is not expected to undermine the dominance of traditional parcel delivery for some time, it will allow market growth for those customers in urban areas who can conveniently pass one of the designated collection points.

Secure unattended delivery companies such as Dynamid provide physical box solutions backed by a database and SMS encrypted messages to allow online confirmation of delivery and collection from each box. These box-type solutions are being targeted at locker banks and high transaction volume households where the number of parcels delivered to each box allows an economic proposition to be developed.

B2C – the challenges

Selecting the right business model, service provider and level of service means understanding the cost and effort involved in generating, capturing and fulfilling each order. It must also be known if and how each order will contribute to the strategic goal of achieving critical mass – how the model will lead to economies of scale that drive Web-based economics.

The bottom line, no matter which model an organization selects, is that they are providing consumers with a new service – often for free – while trying to maintain profit margins. In the old world, the processes of picking items from retail shelves, paying and packaging at counters, loading purchased goods into transport, and delivering them to the home, were carried out by the consumer and were not generally recognized as components of the total shopping cost. With the exception of a segment of consumers who value the convenience factor of home delivery and are prepared to pay a premium for it, today's consumers expect their complete online shopping experience to be price competitive with traditional retailers' shelf prices. High eFulfillment costs will inevitably increase the sustainable price to consumers. On the Internet, though, price differences are completely visible; therefore eFulfillment costs have the potential to blow out and erode an organization's competitive position. In addition, the variable costs associated with product returns are causing many organizations to rethink their operating models.

The choice of eFulfillment business model has enormous implications for the growth and profitability of any product based eCommerce venture and requires a strategic selection that not only recognizes the need for exemplary service performance, but that also delivers a competitive cost structure. Organizations must rigorously question the underlying logic of the business models being considered.

B2B eFulfillment

The dramatic escalation of trade volume through eProcurement solutions – and anticipated volumes through eMarketplaces – has created a B2B opportunity expected to be 10 times that of B2C. As B2B business models evolve there will be an increasing effect on many of the basic trade relationships that exist among supply chain partners. B2B eFulfillment is migrating from long-term one-to-one relationships to fluid many-to-many relationships. Traditional third-party outsourcing arrangements tend to involve complex packages of services provided by one service company – warehousing, picking, shipping, information systems support, labour provision and management all provided under one long-term contract.

Today, these services are increasingly available in smaller components, making it possible for organizations to partner with the perceived best-in-class provider for each capability. For example, Celarix Inc. offers a global package tracking service that can be purchased for a monthly fee. Some software vendors are developing offerings such as warehouse management, inventory management and other supply chain applications on a subscription or transaction fee basis. This virtualization of these basic business processes increases the number of options available to an organization – and can increase the complexity of a company's supply chain, emphasizing the importance of partner alignment in the partner selection process.

B2B eFulfillment models

As the eCommerce landscape evolves, several channels are emerging (see Figure 4.2). The emergence of these channels is driven primarily by the importance of the buyer to the seller and the importance of the seller to the buyer. As such, each of these channels has different fulfillment requirements – some are B2B-specific, others have characteristics that naturally overlap with those required for B2C fulfillment.

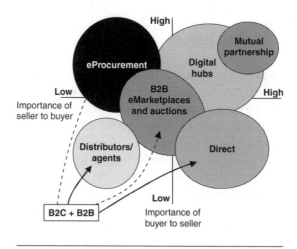

Figure 4.2 eCommerce channels.

Mutual partnership

In a mutual partnership, both the supplier and customer are strategic to each other's business. Automated reordering solutions via EDI or the Internet have been supporting relationships such as these for some time. Similarly, vendor-managed inventory services are most likely to occur in mutual partnerships. The fulfillment solution required to support this channel can include monitoring of material levels at customer locations via telemetry, use of kanban processes to support just-in-time replenishment and replenishment direct to assembly lines. The nature of the physical fulfillment is varied and can include bulk, pallets and tote bins. The emphasis is on high logistics availability to support rapid or just-in-time replenishment.

Examples of mutual partnership relationships include the automatic stock replenishment of industrial gas supply by The BOC Group with major petrochemical customers. In a consumer-orientated sector, Tesco has openly shared its point-of-sales data with several major suppliers. The supply chain planning can then be optimized to minimize cost while ensuring product is always available on the shelf for the customer.

eProcurement

In eProcurement fulfillment, products are frequently supplied from multiple supplier locations to a single customer location. There is a wide range of possible product characteristics that will influence the fulfillment solution required. However, typical MRO items that are replenished via eProcurement solutions are amenable to solutions that consolidate items from multiple sources into single deliveries.

eProcurement is being implemented across a wide spread of businesses by major corporations, including BP, Shell, Dow, Barclays Bank, Merrill Lynch, British Airways, and Pacific Century CyberWorks/Hong Kong Telecom. They are all looking to reduce the complexity of their present transaction systems by implementing the new Web-enabled technology and expect improved transparency of their MRO purchases.

This inevitably leads to consolidation of the supply base with increased aggregation of the catalogues. Office Depot in the USA currently operates a national business-to-business delivery network that includes over 2 000 trucks. Its eProcurement catalogue system is one of the key growth channels for the company, which views fulfillment excellence as one of its differentiators in the market.

Agents and distributors

The two broad categories of fulfillment model in the agent and distributor environment are stocked and stockless. The fulfillment solution required to support the stockless environment is similar to the eProcurement model described above. However, agents and distributors are frequently required to maintain stock in order to meet delivery lead-time requirements. In this case, the fulfillment solution must support the rapid assembly of many small orders from a large number of stocked SKUs. Automated picking can be appropriate in this environment, using sophisticated materials handling systems. History has told us that the automation route should be embarked upon with caution, as inappropriate automation can result in expensive, inflexible solutions that cannot adapt to changing fulfillment needs.

Delivery can be via parcel carriers, specialized third parties or owned fleets. Fulfillment strategies can include: segregation of large and small items into different picking facilities, trunking of orders to trans-shipment locations where they are transferred into smaller vehicles for delivery, holding fast-moving items at local replenishment facilities and slow-moving items centrally, and providing different lead times and availability policies for different products.

The distributor category has some excellent examples. Office Depot, the world's largest seller of office products, aggregates its supplies. The company can fulfil the needs of its various customer types by a daily drop, managing the inventory directly through to making the product available in its own stores for collection. HaysDX guarantees a 100 per cent, next-day service to the legal profession in the UK for documents dropped off before 6 p.m. on the previous evening.

The chemical distributor VW&R, in North America, acts as a service aggregator for multiple chemical suppliers to offer rapid fulfillment and quality service to smaller customers. Distribution channels for MRO items are commonplace, and this sector of the market is being affected by the growth in catalogue suppliers.

Direct channel

In this environment, the eCommerce channel provides an alternative way in which customers can place and track their orders with a supplier. The nature of direct channels range from low-cost direct models in which the emphasis is on reducing the customer's and supplier's transaction costs, to high-value interactive models in which the focus is on increasing the breadth of service provided to the customer.

Where the service and delivery requirements are similar to traditional channels, fulfillment typically will be managed via the existing fulfillment network. Where requirements are significantly different, new fulfillment solutions are required. This is especially the case where a direct channel is being used to bypass existing distributor, wholesaler or retailer channels. Specialist fulfillment services are emerging to meet this need, but are relatively immature and there are no clear leaders.

Several chemicals companies have established direct eCommerce channels. Dow Chemicals offers several products for sale online, including heat transfer fluids and caustic soda. Eastman Chemical Company has established a sophisticated online product ordering and delivery tracking capability for a wide range of products.

B2B eMarketplaces

An enormous range of B2B eMarketplaces are emerging in almost every sector. In the majority of cases, transaction volumes are still very low. However, having established the functionality to introduce buyers and sellers and set transaction prices, many eMarketplaces are now turning their attention to the fulfillment solutions required to support the anticipated transaction volumes.

The ability to confirm availability of fulfillment capacity and provide a price for delivery in real-time is key to supporting B2B eMarketplaces.

There are two key types of exchange where fulfillment is a key feature:

● The first is the virtual exchange where buyers' bids and sellers' asking prices are matched. Virtual exchanges are similar to the NASDAQ in the stock market. One example of this type of trading exchange, ChemConnect, is looking to establish an online fulfillment engine to enable the chemicals to flow between the buyer and seller.

- The second is the online auction, where buyers bid to buy a product or service or sellers quote in response to a request for proposal. The cost of fulfilling the transactions is a key element in the overall cost of purchase. One of the services offered in the i2 Freightmatrix marketplace is to host the logistics capability and cost of service of leading 3PLs. The logistics order can be placed on the Web, and with the track and trace functionality fulfillment can be open.

For marketplaces to be successful they do need to have a digital connected fulfillment engine.

Digital transaction hubs

Digital transaction hubs focus on reducing the cost of integration between buyers and sellers. Participants in a digital transaction hub routinely buy and sell products from one another, and do not require the buyer or seller matching capability provided by eMarketplaces. Several of the world's leading chemicals companies, including Dow, Dupont, BASF and the chemicals division of BP are in the process of developing a digital transaction hub called Elemica. Once established, members will achieve integration with each other by integrating into the hub itself.

Digital transaction hubs provide fascinating opportunities for innovation in fulfillment. The hubs will allow member companies to collectively outsource fulfillment activities that they do not consider to be core or differentiating. This will allow the member companies to collectively achieve economies of scale that individually would have been impossible. Conversely, a provider of fulfillment services need only integrate into the digital hub to provide services to the member companies. This will enable the provision of value-added services such as track-and-trace and supply chain flow management at a cost and on a scale that previously has not been possible.

B2B eFulfillment solutions

Selecting an eFulfillment solution must incorporate consideration of factors such as speed of implementation, degree of customer contact and operational control, reliability and flexibility of service and initial and ongoing costs. One of the most critical considerations for an organization is to select the business solution that best combines their capabilities with those of its chosen supply chain partners. Models available include spot shipment, using consolidators, employing a

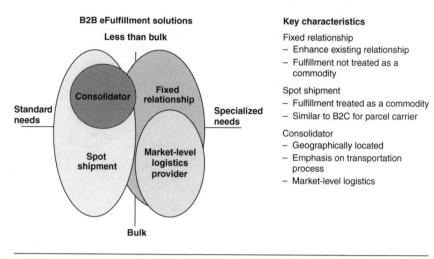

Figure 4.3 B2B eFulfillment solution landscape.

market-level logistics services provider, and developing fixed relationships. The advantages of each model vary, depending upon the degree of specialization required, and whether shipments are bulk or less-than bulk (see Figure 4.3).

Fixed relationship

With a fixed relationship solution, fulfillment service providers and users maintain a long-term relationship. This is most likely to be appropriate where there is a degree of specialization in the fulfillment service required. For example, where there are only one or two service providers that have the required capability, the value of a long-term relationship outweighs the benefits of finding the lowest spot prices day to day. This model is particularly relevant where an investment is required on the part of the service provider to meet the specific needs of the user. In this kind of relationship, there will typically be a commitment by each side to a certain volume of business. There is substantial opportunity for the service provider to differentiate itself via value-added services.

Spot shipment

Spot shipments are applicable in environments where the fulfillment requirements are relatively standard. Parcel delivery, full truck dry goods haulage and container shipment are examples of relatively standardized services that are amenable to spot shipment. Spot shipments are also relevant where the fulfillment requirements of a user are varied and unpredictable, such as in project-based activities like construction. Spot shipment solutions cover a wide range of physical fulfillment activities. They are by their nature amenable to being provided via either public or private fulfillment exchanges.

Consolidator

In a standardized or less-than-bulk environment, there is an opportunity for service providers to develop consolidation services. A consolidator collects or receives products from multiple suppliers and consolidates them via a cross-dock operation into single deliveries for business customers or consumers. This could potentially reduce overall fulfillment cost and increase convenience for the customer. It could also eliminate the need for a centralized stockholding to serve B2B or B2C channels.

Market-level logistics provider

Where fulfillment needs are shared across an industry or market, there will be an opportunity to establish market-level solutions. Such solutions will emerge in industries where provision of the fulfillment service requires substantial capital investment that can be leveraged across multiple users. Such solutions already exist in some sectors, for example, retail petroleum distribution to gas stations. However, a number of factors have limited the development of market level solutions, including a limited willingness of competitors to collaborate, concerns about the security of customer information and sales volumes between competitors, the challenges of integrating with multiple and varied systems and, concerns that a market-level solution would commoditize fulfillment services.

These challenges are being overcome in some industries. For example, the establishment of the chemical companies' digital hub, Elemica, signals a willingness to collaborate within a fulfillment network and develop industry solutions. The creation of digital hubs creates the opportunity to provide services not just to individual users but to the network as a whole.

New eFulfillment business models

Enabling technologies that provide the synchronization of information and processes across trading partners are central to the success of eFulfillment capabilities. It used to be the case that the fulfillment solution landscape was relatively straightforward. Services were relatively easy to classify – third-party logistics, haulage, freight forwarding, parcel carrier etc. However, the emergence of supply chain networks creates the opportunity for completely new kinds of service – and several new solutions have emerged (see Figure 4.4). Freight marketplaces match buyers and sellers and can provide a mechanism to set the price at which a service is provided. This includes exchanges, auctions and reverse auctions. Infomediaries provide information through a network that supports synchronized decision making at the operational, tactical and strategic levels. And finally, flow management services manage the flow of transactions through a network. An extension of this capability is the provision of supply chain planning services to network participants.

Freight marketplaces

Marketplaces can be created for a variety of fulfillment service offerings – freight marketplaces have developed the most rapidly. This space is

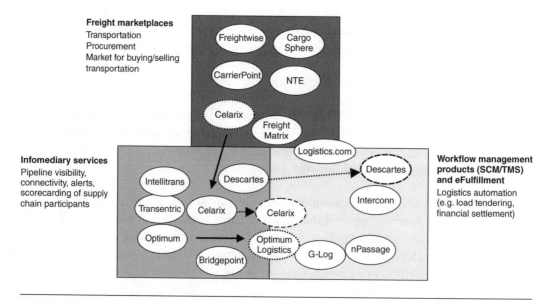

Figure 4.4 New eFulfillment business models.

currently very crowded, with in excess of 150 such marketplaces operating globally. Most focus on a particular transport mode, and some take an industry or geographic focus. The marketplaces provide a variety of value-added services, which largely fall into the infomediary or workflow management categories described below. Spot fulfillment needs can be fulfilled and marginal capacity sold via public exchanges.

Freight marketplaces fall into two major categories – public and private. Public marketplaces are open to any carrier or shipper that wishes to participate. Private marketplaces are restricted to member providers and users. Private marketplaces are typically constructed around a specific synergy opportunity and tend to be focused on specific modes and geographies. Although public marketplaces are currently more prevalent, private marketplaces could gain traction rapidly because they are constructed around a specific synergy opportunity, are focused on a specific requirement, service providers are pre-vetted and known by users and all participants are committed to the success of the exchange.

CASE FILE

eLogistics operates an online road freight procurement exchange that matches shipper's road freight requirements with carriers or truck owners that are able to move their load. eLogistics Freight™ went live in October 2000 and has seen a rapid increase in the number of shippers and carriers using the system.

eLogistics also now has the ability through eLogistics Integrator™ to seamlessly integrate logistics fulfillment into B2B eMarketplaces. They have completed, or are in the process of completing, alliances with metals procurement exchanges, construction exchanges and paper exchanges.

The challenge for B2B eMarketplaces is to integrate the logistics element with the procurement element. eLogistics Integrator™ means the company can now link a trading exchange, through the eLogistics Freight™ platform to the resource management systems of some of Europe's leading logistics service providers as well as the Web browsers of thousands of smaller carriers.

Infomediaries

'Infomediary' describes a broad range of services that provide information to a community of users, including specific track-and-trace information, industry trends, prices and news-feeds.

In the context of eFulfillment, infomediary services tend to focus primarily on the provision of track-and-trace information. The provision of such information is no trivial matter. Not only does the information need to be collated and presented in a format customized to the needs of the user – arguably the easier component – it also needs to be gathered at the point that the information is created, be that in loading, in transit, or at the delivery point. The practical challenges to capturing and recording high-quality and complete information at this level are great. Traditional approaches, such as keyboard entry and barcode scanning, are now being complemented by technologies such as GPS (satellite positioning), mobile phone location and radio tagging. As these technologies mature, we can expect to see the range of infomediary services continue to develop, enabling shippers to achieve new levels of supply chain integration.

CASE FILE

Celarix enables the flow of supply-chain information between networks of companies and provides a foundation for collaboration. With the increase in collaboration between companies and the need for visibility of partner inventories and systems, internal and external integration requirements have increased tremendously. Existing tools allow companies to transmit data between one another. However, the key to collaborative success is to transform the data into information that can be used in daily business processes and acted upon by each member of the network.

Celarix offers a suite of commerce, management and intelligence services that enable global organizations to collaborate with their suppliers, customers, financial institutions, transportation providers and brokers by transforming supply chain data into relevant information. To obtain the maximum value from these applications, Celarix provides a flow of high-quality data, regardless of the source systems, data formats and modes of transmission.

Working with customers and partners to agree on business rules, messaging standards and to coordinate the flow of data is difficult and time-consuming, but are critical components of any supply-chain collaboration programme. Celarix solutions integrate with customers' and partners' internal transportation and warehouse management systems and procurement, customer service and advanced planning and scheduling applications. Celarix also integrates with external factories, distributors and value-added resellers (VARs) as well as over 150 leading freight forwarders, consolidators, trucking companies, ocean lines, airlines and brokers. Once information is contained within the Celarix applications, customers are able to improve service levels, lower inventories (up and down the supply chain), position inventories at the point of demand and increase both organizational and network performance.

Flow management solutions

Several solutions are emerging that focus on the management of supply chain flows and transactions. These services enable firms to outsource not just fulfillment execution and scheduling, but also the management and optimization of these activities.

Fourth-Party Logistics™[1] (4PL) providers are emerging to manage the flow optimizations across multi-company product flows. Ryder Logistics and others have offered such a service which can lead to cost savings of 5–15 per cent of logistics costs, depending on the market sector.

The critical ingredient is for the 4PL to access common transaction data flows from the participating companies. This source will be available from some of the marketplaces being developed, and as such we foresee a growth in 4PL activity in the near future.

CASE FILE

Yantra provides technology solutions that help businesses manage order management transactions across complex, multi-partner 'extraprises'. The solution provides end-to-end order life cycle visibility through a given supply chain network of partners. Additionally, custom business logic can be developed to identify business exception conditions that alerts management to take action via Web-enabled communication tools.

'PureEcommerce', Yantra's transaction management engine, enables integration and execution across an extended enterprise of trading partners. Designed specifically for entraprise environments, PureEcommerce provides end-to-end order management visibility throughout the entire order life cycle. For example, PureEcommerce serves as the system of record for all extraprise order management transactions. It allows clients to operate as a single order management system in a complex extraprise environment.

Conclusion

B2C eFulfillment has had to address significant complexities, including large numbers of relatively low-value orders, call centre integration and management of returns. B2B adds a whole new range of complexities including a variety of direct and indirect products, specialized product

---

1 Fourth-Party Logistics™ (4PL) is a registered trade mark of Accenture.

handling, a range of ordering mechanisms and the full spectrum of parcel sizes from boxes of pencils to sea vessels full of petrochemicals.

What is clear is that supply chain synchronization will be a key feature of B2B. Enabled by emerging supply chain visibility, workflow management and decision support tools, supply chain synchronization integrates the supply chain across a network of participants. Success in this environment will rely upon the development of warehousing, transport and customer service solutions designed to deliver operational excellence in fulfillment. So too, the establishment of new kinds of relationships between supply chain participants – such as increased collaboration between service providers and users, win–win commercial arrangements and a true understanding of core competencies and reassessment of activities that can be outsourced – will be a critical feature of new fulfillment approaches.

An integrated fulfillment approach combines eEconomy opportunities with a 'back to basics' focus on operational excellence. Ultimately, it is those companies that can deliver their products to their customers in time, at the right quality and at the right cost through the new eCommerce channels which will be successful.

5 The eDesign and eManufacturing challenge

Introduction

eCommerce is also having a profound impact on today's manufacturing environment. The Internet and associated new technologies are increasing the demands placed on design and manufacturing organizations as well as providing the mechanisms required to meet these new demands. Industries everywhere are focusing on lower cost and higher speed solutions for developing and manufacturing their products, as well as new approaches to managing inter-company relationships. Product life cycles are shortening and, in turn, amplifying the need for agility and speed-to-market. Indeed, speed-to-market, a measure of how quickly a new product can be designed, tested and an initial number of units produced, and time-to-volume, a measure of how quickly new product manufacturing can reach full capacity to meet demand, have become key competitive capabilities.

In this chapter, we explore some of the key drivers of the changes being experienced in design and manufacturing, and the key developments that we see emerging as a result of these drivers. The specific questions we address in this chapter are:

- What trends are requiring design and manufacturing organizations to change?
- How are design and manufacturing organizations responding to these trends?

Drivers of change

Few companies are untouched by the challenges – and the benefits – of globalization and eCommerce. An increase in industry 'clockspeed' or speed of evolution translates into consumers and trading partners becoming more demanding in terms of timing, quality and segment or customer-specific product preferences and requirements. Innovations are being developed and circulated across industries and geographies at a faster rate.

The ability to reach global customers similarly is adding to the complexity and competitive intensity of today's manufacturing environment. Customers are more aware of, and have better access to, a range of manufacturing alternatives and are no longer willing to accept 'one size fits all' products that only partially meet their requirements. And so, manufacturers are required to be more nimble and responsive to customer requirements.

Many organizations have turned to offshore manufacturing facilities to reduce costs and improve performance. A shift to manufacturing in low-cost countries is supported by Internet-enabled real-time global communications and collaboration capabilities. However, traditional issues – such as availability of skilled labour, quality assurance, existence of appropriate infrastructure, and the extension of lead times in the supply chain – still remain and must be squarely addressed by a manufacturer's organizational and operational strategies.

Likewise, traditional original equipment manufacturers (OEMs) are increasingly choosing business models in which they outsource or subcontract much or all of their manufacturing operations in order to concentrate on core competencies such as customer relationship management, flow management, research and development and product and manufacturing process design and specification. The sharing of design and manufacturing information across organizational boundaries is therefore an increasingly critical capability requirement.

In the area of product design, many companies are growing increasingly dissatisfied with the returns on investment in research and development (R&D). Accelerating industry clockspeed is pushing new product development into shorter time frames, decreasing product life cycles and reducing the time available to recoup investment costs. Organizations and whole industries are facing rising R&D costs – with a decreasing return on investment. Two industries for which this is particularly true and threatening are the pharmaceuticals and electronics and high-tech industries. Rising costs and availability of new technologies are leading to new solutions for product design.

CASE FILE

Since 1993, the research and development spend of the top 20 pharmaceutical companies has more than doubled, and is forecast to double again by 2005. During the 1980s, R&D spend in the pharmaceuticals industry, which was then growing at around 11 per cent per annum, represented about 10 per cent of turnover. In recent years, the R&D figure has grown to more than 17 per cent of sales, with turnover growth slowing to about 6 per cent per annum.

The fact that there are more products in the 'development funnel' in part accounts for the rise in R&D expenditure. However, research by the Centre for Medicines Research (CMR) indicates that, for the companies polled, average output of new active substances remains at less than one per year.

If the top 20 pharmaceutical companies are to achieve their target of 7 per cent annual growth, the average company sales from new products between now and 2005 must be around $28.9 billion, equating to at least 24–34 new products, each earning between $1 billion and $1.45 billion. Average revenues per product are currently just $265 million per annum. If these companies increase their R&D expenditure in line with historic growth rates, the total shareholder returns will fall from nearly 30 to 10 per cent per annum. If they spend more under the same conditions, total shareholder returns will crash.

In order to avoid takeovers and forced boxing into niche markets, organizations must cut both R&D costs and lead times, generate additional sales revenues with best-seller products or enter new markets. Analysts predict there could be as few as 13 industry giants that will still be industry giants by the year 2005.

Industry consolidation of design and engineering expertise is an increasingly popular approach to maximizing skills and intellectual capital. Engineering design centres of excellence are emerging in select engineering fields, perhaps most notably in the automotive industry. As large automotive companies widen their product portfolio, they have begun to leverage specialist design outfits to supplement their skills. This is not 'outsourcing' of design expertise – another approach adopted by some organizations – but the use of specialist skills to plug gaps in knowledge or to aid the faster introduction of new models.

These change drivers, together with the uncertainty of product supply – which comes from the ongoing reliance on external suppliers and the emergence of competitors from non-traditional sectors – are creating a dynamic in the manufacturing sector that is far more demanding than

CASE FILE

Lotus Engineering, a UK-based company, is a centre of excellence in the automotive industry. From a humble beginning of designing cheap open-top sports cars in the 1960s, through a dubious involvement in the ill-fated DeLorean venture, Lotus is now seen as one of the leading chassis and drivetrain engineering companies in the world. Most recently they have worked for General Motors in designing the VX220, a two-seater sports car that is about to launch into full production in Europe. Lotus's capabilities are now being used in the USA with extensive engineering design and test facilities in Ann Arbor, Michigan. Lotus has also managed much of the design work for the Honda Civic.

Ricardo Engineering, also in the UK, is another example of an engineering centre of excellence. This company specializes in the design and selected manufacture of transmission components. They work with many of the leading high-end sports car manufacturers and have also recently supported BMW in the integration of the engine and transmission into the new BMW Mini automobile.

we have experienced in the past. Original equipment manufacturers are developing several options to move themselves and their partners to the new required levels of product design and manufacturing capability.

Design and manufacturing responses

Customer-driven eDesign and eManufacture

Customers are fast becoming closer to – indeed part of – the design process. The data that originally sat inside computer-aided design (CAD) systems is being restructured with a consumer-friendly Web site front end, allowing customers to tailor their own products. Online shoppers can generate a product model and choose and change specific features. A number of leading companies have already embarked on offerings that incorporate this interactive capability. Sony Corporation, for example, allows its online customers to choose the operator controls on its PlayStation2 joystick before making a purchase. Dell and Compaq allow customers to configure their PC components online, and many automotive manufacturers are beginning to allow customers to configure the features of the car they want rather than limiting them to stock products. Taking this a step closer to the design process, rather than just the configuration process, Fiat allowed a group of its key customers to help in the conceptual design of the next-generation Fiat Punto, and Microsoft cleverly used its customers to assist in the development of Windows 2000. The result is products that are more closely tailored to what customers want.

CASE FILE

The Fiat car company took advantage of the connectivity capabilities of the Internet when it conducted a Web-based survey to evaluate its customers' needs for the next-generation Fiat Punto. The 3 000 survey participants were from Fiat's key target segment. The customers could design a car on-screen by selecting from various styles and features. The software used not only captured the results, but also tracked the steps that customers took when evaluating and selecting options, thus giving the company further insight into the priority placed on certain criteria.

Fiat received 30 000 pages of data – valuable information that directly influenced styling and concept designs. The total cost to Fiat was significantly less than it would have been to run a series of focus groups, and the quality of the knowledge gained far exceeds that gleaned in previous focus group activities.

CASE FILE

Microsoft is well accustomed to using its own highly technical people to create a constant feedback loop between old and new product versions. More recently, the company used 650 000 of its customers as a live test-bed for its beta version of Microsoft Windows 2000. Customers provided ideas on changing some product features and clearing glitches from previous versions. In addition to providing valuable feedback, some customers even paid Microsoft for the privilege of testing the product in an effort to understand how Windows 2000 could create value for their own businesses. It is estimated that the customers' contribution to co-developing Windows 2000 was worth over $500 million in time, effort and fees. The intangible benefits associated with the quality of the input and consequential impact on the success of the product are naturally more difficult to assess; suffice to say they are considerable.

Leading organizations also are using customers as virtual help teams that answer queries and share user tips. These direct links to customer communities provide product companies with live test-beds – true market environments free of the biases of traditional focus groups, and a mechanism that helps facilitate personalized experiences with customers. Such systems can also be used to ask respondents what they like or dislike about products offered by competitors, and that feedback can be incorporated into new product design and development.

Virtual Internet communities that are created through the use of product design and development portals provide companies with the opportunity not only to leverage their customers' ideas in product design, but also to leverage their customers' experiences in promoting their products. Organizations are finding that many customers are willing to communicate their experiences about products and services through discussion portals. Given that this information may be visible to other potential customers, it serves as an efficient advertising mechanism with a powerful message. In addition to providing product feedback information, there are an increasing number of instances where customers offer direct product choice advice to other potential customers. For the product organization, this has provided access to valuable information such as which product features customers are particularly happy with and should therefore be continued in later models and upgrades. Previously, this was an almost unsourced information area, essentially neglected in the old 'complaints' call centre models for customer feedback.

Clearly, product organizations incorporating such visibility and information sharing through their portals must be committed to meeting their customers' demands in order to minimize the risk of bad press. Whether visible to other customers and consumers or not, the negative feedback on product design and performance also provides the organization with valuable and critical information. The direct feedback capability enables organizations to track and quantify service and support costs faster and with a much higher degree of accuracy, a task previously fraught with inaccuracies and delays given the range and often disjointed nature of the feedback, service and support mechanisms and channels. Clever companies are encouraging customers to communicate in this way and are feeding the information directly to where corrective action can be taken.

Greater collaboration between design and manufacturing engineers

Incorporating manufacturing considerations and ideas into the product design and development process is not a new concept. However, until the emergence of Internet-enabled technologies that allow companies to

CASE FILE

Jaguar has direct links from its showrooms and workshops to the final quality assurance station on the production line. Any problems experienced by customers or solved in workshops are immediately fed back to the production line, enabling timely focus on the cause of faults. The quality insurance inspector can discuss the problem immediately with the workshop supervisor so that the issue does not get lost in paperwork.

collaborate more efficiently and effectively, both internally and across organizational, geographical and cultural boundaries, many approaches to design-for-manufacture, design-for-assembly, design-for-service and design-for-disassembly and recycling have not been affordable and sustainable options for manufacturers. Today, manufacturers are just beginning to adopt product development practices and tools that will help them to deliver innovative products at Internet speed.

Product design and development portals offer a single point for access and control to all parties and processes involved in product development projects. They allow universal, low-cost, real-time linkage and communications – via instant messaging and Web conferencing – between key suppliers, original equipment manufacturers, contract manufacturers, engineers, marketers, designers and even customers. Organizations already using design and development portals are rolling out higher quality, more customer-specific products in less time and without many of the interdepartmental issues that are typical of more traditional design processes. National Semiconductor, for example, allows its customers and supply chain partners to collaborate through product design portals from the early stages of circuit design, saving time and money and creating more appropriate end products.

The most successful design and development portals will bring together news and information, parts catalogues and data sheets, downloadable CAD drawings and models, training courses, collaborative software, and much more. Along with the product development projects, effective portals will enable engineering activity to occur concurrently between component and equipment suppliers. They provide the project manager and members of a cross-functional design team the ability to manage the design collaboratively from a very early stage. Given that around 80–90 per cent of all cost and quality decisions are made within the initial 5 per cent of the product life cycle, the impact of this up-front collaboration is disproportionately high.

CASE FILE

Adaptec migrated a range of business processes online, including collaborative engineering. It incorporated the development of capabilities such as real-time information sharing from the early stages of prototype design, development and testing and automated work-in-progress updates and forecast sharing. Adaptec has been able to recapture its initial investment many times over; manufacturing cycle times have been reduced by up to 50 per cent and inventories reduced by 25 per cent.

Three-dimensional (3D) modelling technology is similarly improving product design processes. While it is true that most organizations still use two-dimensional (2D) wire frame models for product design on the Web, the uptake of 3D modelling technology is now occurring at an impressive rate, with many benefits. Engineers can visually share the internal and external characteristics and features of a component or assembly so that everyone involved in the process, from suppliers to customers, can see what the proposed product will look like. Engineers can also provide product density information so that the model has weight and mass, and can even calculate a number of the product's physical properties. 3D models will identify interference with other electronically generated parts automatically, thereby improving the quality and speed of the design process. They can interact with production tooling, which is also designed using 3D techniques so that the profiles will be an exact fit.

Leading manufacturing organizations and their partners are now using 3D electronic modelling in the early concept-testing phase of the design and development process – testing products that exist solely in the computers of their engineers. This enables the identification of design faults, alternatives and improvement opportunities prior to production, and is therefore leading to the creation of higher-quality products with shorter cycle times at a lower overall cost.

Beyond 3D modelling, virtual reality has the potential to unlock better, faster and cheaper product design. Clear benefits exist over traditional physical 3D modelling, including accessibility, flexibility and overall

CASE FILE

Matrix Automation Group, an American organization that designs and fabricates custom automation machinery, has moved from a 2D design package to a 3D modelling design software system called SolidWorks. The impact on time-to-market has been dramatic. The process of turning a layout into original detailed drawings has decreased by a ratio of 8:1 and the software has eliminated the need to rework all the individual drawings when there are design changes. Further, design faults and mistakes that were difficult and slow to detect through the old process of detailed examination of 2D drawings can now be detected automatically in a matter of seconds. This has resulted in additional days saved through reduction in examination time and through the elimination of costly rework due to undetected product faults progressing to the next phase in the development process. Matrix has seen its production costs decrease and its on-time delivery performance increase significantly.

cost. An area of business particularly suited to the use of virtual reality is the customer interface to enable customers to virtually 'test run' products. This has the associated benefits of simultaneous multiple user access, extended and instantaneous geographic coverage and visibility of alternative components and options. Training is another area where companies can use virtual reality to maximize effectiveness and minimize risk. In a manufacturing environment, for example, employees can practice new operational procedures using a virtual assembly prior to using the physical system.

CASE FILE

Sponsored by NASA, The Virtual Collaborative Clinic will bring medical clinics to patients via high-performance networks to enable rural and under-served communities to share in the benefits of advanced medical technology. The clinic will enable physicians and scientists to use multimedia in a distributed network to plan and practise delicate operations in virtual space. Digital libraries of virtual patients will allow doctors to share information about rare procedures and provide a powerful teaching tool for future generations of physicians. Remote hospitals will be able to access the knowledge, skills and techniques of larger institutions. Many government, medical and academic bodies are involved in this pioneering effort, together with a number of industry players, including Cisco Systems, Intel and MCI. The ultimate long-term goal is to extend the Virtual Collaborative Clinic technology into space to service future astronauts on the International Space Station and beyond.

Virtual reality is now close to being available on the desktop computers of engineers and designers to help organizations reduce their dependence on physical prototypes. Consider a typical prototype of a car. A physical prototype can take around 13 to 26 weeks to build, days or weeks for each set of alterations in the design iteration process and hundreds of thousand dollars for development time and materials. A virtual prototype will reduce the build time by at least 60–70 per cent, and alterations can be made in just minutes. For the average manufacturer operating in the eEconomy, the recovery of initial spend on the new virtual reality technology and its implementation is quickly, if not immediately, recoverable given the increasing demand for new product introductions and upgrades.

A virtual prototype, however, has limitations that must be given consideration. It must incorporate manufacturing process knowledge, the sequence of fit and how the parts are manoeuvred into position without fouling other parts. In addition, the virtual data is nominal, that is, plus or minus zero tolerance, not the actual manufacturing allowed tolerances that will exist in reality. Therefore designers must consider, for example, how a physical car will react in crash tests when built through manufacturing processes that, in reality, will not be perfect every time as they would be in a virtual model.

CASE FILE

Engineers at General Motors are using virtual reality tools to simulate how a new vehicle design will withstand various crash scenarios; the vehicle structure performance; and the occupant protection performance. With virtual reality, they are evaluating the central systems of a vehicle and the performance of components, subsystems, and the vehicle as whole.

The construction industry is one of the most suitable industries for the use of virtual reality modelling. Geographically dispersed architects, engineers, designers, potential investors and end-users can collaborate in virtual space to design a building, a ship, a town or a city. No scale model is required when a virtual prototype exists. The virtual prototype can be entered, walked through and examined in detail. It can be exposed to simulated elements and scenarios to examine its capacity and ability to withstand an endless variety of situations. It can be redesigned, adapted and improved on any scale in a very short amount of time and at a dramatically reduced cost – it can be rebuilt even before one brick is placed on top of another.

While virtual reality systems vary depending on the targeted application, they all consist of four basic elements:

1. Some form of inexpensive, stereoscopic, lightweight visualization eye-wear that can be tracked by the computer in order to achieve the proper 3D perspective.
2. A 3D input device, other than a mouse and a keyboard, such as a data glove or something that would enable the user to work with the data in a more natural manner.
3. The specialized software.
4. A reasonably powerful computer.

Virtual design environments provide users with a sense of being immersed in the three-dimensional computer-generated world, and the ability to interact with the computer models in a manner that closely resembles real life.

Although still in its infancy, we are beginning to see the use of virtual reality in the design, prototyping and testing of products. Design and engineering teams are coming together in virtual laboratories, workrooms or studios to build virtual prototypes that can be explored and tested under a multitude of virtually applied scenarios in order to detect faults, analyse performance and identify design improvement opportunities, all before the first step in the manufacturing or construction process.

Unlocking the value of research and development investments

Development portals hosted by eMarketplaces and unbiased third parties are becoming the core of product design and development in the eEconomy, changing approaches to patents and other intellectual properties. Rather than back-shelving their intellectual assets or relying upon heavily guarded independent negotiations for their exchange, corporate giants are now willingly unveiling the precious fruits of billions of dollars worth of research and development spend. These

CASE FILE

yet2.com, a privately held corporation, is a global forum for technology and intellectual property exchange on the Internet. It has already enlisted many leading corporations and government agencies from a vast assortment of fields to pledge their technology assets. yet2.com's intention is to create the most efficient method for streamlining research and development and extracting value from intellectual assets. Companies joining yet2.com commit to offering their technology for license or sale exclusively on yet2.com's Web site. To date, the signed-up companies collectively make-up more than 10 per cent of global spending on research and development.

yet2.com's impressive list of global forum members includes: Minnesota Mining & Manufacturing, Boeing, Dow Chemical, DuPont, Ford, Honeywell International, Monsanto, Polaroid, Procter & Gamble, Rockwell International, SAIC, TRW, Agfa-Gevaert Group, BASF, British Telecommunications, Ciba Specialty Chemicals, Philips Electronics, Shell Global Solutions, Bayer, Siemens AG, Porsche, Bosch, Asahi Glass Company, Denso Corporation, Fuji Photo Film Company, Kao Corporation, Mitsubishi Chemical Corporation, Mitsui Chemicals, Sumitomo Electric Industries, Toray Industries, Toshiba and Toyota.

eMarketplaces are facilitating the commercialization of innovations and ideas. Emerging with a wide variety of new value propositions, some are adopting vertical industry models, while others, such as yet2.com, are clearly horizontal plays.

What yet2.com and others have recognized is the opportunity that lays dormant in the massive investments in research and development that never become actual products. While the original developers of these technologies and intellectual properties may not have found a use for their developments, companies like yet2.com believe that others will. The value proposition that these eMarketers of intellectual property offer to proprietors is generally a combination of a carefully managed means by which they can extract real value from their undervalued or unused technology and innovations, and access to previously undisclosed technologies from other leading research and development organizations. In most cases, potential buyers are able to search the site for intellectual property and solutions that are relevant to their own organizations – solutions that are already available and that entail a significantly lower cost than that which would be incurred if they had to invest in their own research and development programmes.

Alternative business models for neutral, third-party, intellectual capital eMarketplaces include those that focus on access to rich skills and

CASE FILE

TechEx provides a facilitated concentration of research and licensing professionals to the biomedical industry. It offers an online forum where approved members can introduce technology and intellectual property that they wish to make available for partnering. TechEx has three types of participants:

1. Qualified licensing professionals from research institutions.
2. Qualified corporate licensing professionals capable of bringing early stage inventions to market or otherwise providing significant value-added development.
3. Qualified venture capitalists capable of providing financial assistance to commercialization efforts.

The corporate licensing professionals and venture capitalists are required to establish a confidential and secure interest profile describing the licensing opportunities that would be of interest to them. The corporations with the technology for out-licence and the research institutions provide descriptions of their technologies to the extent that they consider non-confidential. The role of TechEx is to match the opportunities with the interests in order to complete the arrangement.

knowledge in a specific area rather than on a broad range of existing technologies, and those, such as *simply*engineering, that are incorporating and encouraging real-time collaboration on the Internet and the optimization of (engineering) capabilities both within and across industries. Some will attempt to offer it all.

Information-based eMarketplaces have the potential to provide enormous opportunities and benefits. They will provide all companies, irrespective of their size and historical research and development spend, with the opportunity to compete in an increasingly time-critical environment. They will enable industry and cross-industry groups alike to optimize the use of their engineers, designers, project managers, innovations, and other stores of intellectual capital.

Reinventing manufacturing processes and organization

Shrinking product life cycles and increasing competitive intensity have been fuelling the need for manufacturing agility. To establish and maintain a competitive edge in the eEconomy, manufacturers must now, more than

CASE FILE

*simply*engineering intends to drive value to a company by adding high value to activities like engineering productivity and by liberating previously hidden intellectual property. It plans to make simple to sophisticated modelling software accessible and usable online. The company will broker, within a multi-dimensional index structure, white papers, chat rooms, conferences, Webcasts, online training, and access to specialized services and products. Given that approximately 30 per cent of an engineer's time is spent documenting and communicating, *simply*engineering will bring collaboration to the fore. It will establish *de facto* standards for engineering modelling and analysis activities, and will partner with organizations to build linkages into corporate and Web-sourced project workbenches and workflow engines.

ever, streamline their internal and external processes to maximize the speed of new product introduction and eliminate the gamut of delays and quality problems that typically occur when it comes to escalating production volume. The required degree of process improvement and acceleration, combined with the new tasks of managing relationships with contract manufacturers, presents new levels of complexity.

To meet the ever-increasing demands of customers and partners, all organizations involved in the manufacturing process, from the OEM to the commodity suppliers, must have the capability to keep track of the latest versions and iterations of the products being built. They must always know what is required of them, and whether the right parts will be available at the right place at the right time in the right quantities to meet the production demands. The use of contract manufacturers and the outsourcing of component design to suppliers clearly carries new control risks that are best managed through tightly integrated communication and collaboration channels.

The same kind of increasing complexity and need for rapid communication and collaboration also exists for non-outsourcing OEMs. These companies must coordinate manufacturing processes and aggregate purchasing processes across their own multiple manufacturing sites. Therefore, in addition to product design and development portals, we expect to see a proliferation of manufacturing project portals and, given that the majority of manufacturing projects are associated with some new product introduction, integrated product and process design portals. Through these, all parties will have access to up-to-date information and the opportunity to share ideas and issues with their supply chain partners in real time.

The ability of value chain network partners to capture and share information across traditional organizational boundaries in real time is fundamentally changing the speed and quality of manufacturing decision-making. A high degree of collaboration in the product design and supply chain operations planning phases will result in quality decisions that will dramatically improve customer service levels and quality and will drive inventory, time and cost out of the system. In fact, companies will be able to operate with fewer manufacturing facilities and support wider, global markets.

Effective use of the Internet design environment results in the design of supply chain 'friendly' products. Put simply, the benefits of efficient, collaborative eDesign practices are compounded further downstream in the supply chain. Effective eDesign will increase manufacturing flexibility and minimize design and manufacturing complexities that would otherwise drive supply chain inefficiencies and cost. It will help avoid downstream issues in production, logistics and service parts, and reduce manufacturing cycle time.

Improved manufacturing responsiveness may be achieved through a number of manufacturing approaches, including lean manufacturing, smaller lot sizes, configure-to-order, build-to-order, postponement-delay differentiation and outsourcing. Configure or build-to-order and postponement capabilities are entirely dependent on effective design-for-manufacturing capabilities. Successful eDesign and eManufacturing require the joint development of product and process strategies that meet diverse customer needs. These approaches optimize the use of existing order lead-times to produce product at optimum time in the production schedule. Obsolescence and issues associated with short product life cycles are minimized as the decision on the final product configuration is driven by actual customer demand.

Players in the electronics and high-tech industries, such as Dell Computer, have gravitated to build-to-order strategies. This is primarily due to the sub-assembly-oriented nature of the products and operations in the industry and the new online customer configuration capabilities that have been widely adopted by these technologically sophisticated organizations. Many other industries are trying to follow suit.

CASE FILE

The magazine and catalogue printing industry has adopted build-to-order strategies by incorporating information technologies to create personalized catalogues that bundle products that match unique end-customer specifications. In this industry, high-speed digital pre-press systems and data manipulation allow high-volume printing lines to print individual catalogues as part of a long, high-volume print run.

White good manufacturers GE, Maytag and Whirlpool are analysing the impact that increased information visibility and Internet-related orders will have on their capital-intensive businesses. The concept of build-to-order may mean chopping the massive production lines into flexible cells that can adapt to the mass customization requirements that Internet one-to-one marketing may offer. Smaller runs based on actual orders will potentially reduce downstream inventories, but could cause considerable rework of upstream planning and execution processes. The real-time capture of demand signals may facilitate shifts from traditional manufacturing methods.

The objective, of course, is to ensure that the product the customer wants is delivered as quickly and cost effectively as possible. Typical programmes that companies are employing are expert configuration, cellular manufacturing, cross-training of workers, and performing final product configuration as close as possible to the customer. Some even have the final differentiation processes completed in the distribution network.

Collaborative manufacturing: OEMs and contract manufacturers

Original equipment manufacturers are increasingly becoming virtual manufacturers. They outsource product sub-assembly, and component manufacturing and assembly to tier-one suppliers. On the surface, the transfer of manufacturing assets to the tier-one suppliers leverages their core competency – manufacturing – thus enabling the OEMs to focus their efforts on brand management, customer management, and product innovation.

Below the surface, however, the financially driven trend to outsourcing can adversely affect the supply chain. Typically, outsourcing does not streamline or speed up or simplify the supply chain, rather, it makes it more complex. Interfaces between OEM and tier-one suppliers increase, information becomes more diffuse, and plants and facilities have new owners. The long-term success of the outsourcing model, therefore, depends upon effective collaboration between not only the OEM and the contract manufacturer, but also the contract manufacturer's suppliers and the OEM's customers. Many organizations manage their outsourcing relationships as transactional rather than collaborative relationships. In this model, communication between partners and across business processes is essentially one-way: data, such as forecast information and product content changes, flows from the OEM to the tier-one supplier with little or no advanced information about future needs. Because information flows only one way, the tier-one supplier is always reacting, never planning, and does not give the OEM any information on material or capacity availability – or even confirm that it has the most recent release of product information.

Outsourcing adds new complexity to the process of assuring that information is correct and consistent in the ERP/MRP system. The transfer of product information, such as bills of material and drawings, across the new boundary requires a whole new process to convert the data that the tier-one supplier can use and to validate part numbers and approved vendor lists. In an environment of shrinking product and component life cycles, supply changes are common. All these changes must be converted and validated – a process that is often manual and therefore time-consuming and susceptible to error. The result: lengthy time-to-market and excess and obsolete inventory.

In this transactional supply chain, both the OEM and the tier-one supplier tend to add additional resources and layers of management at the point of interface to manage the new challenges. Potentially, these costs outweigh the financial benefits of the outsourcing arrangement. OEMs that have recently outsourced their manufacturing capacity to tier-one suppliers have already realized the financial benefits associated with reduced assets – by establishing a transactional relationship, these OEMs take on a significant competitive risk.

The first step toward collaboration is to enable two-way communication. For example, let's say an OEM places an order with the tier-one supplier for a quantity of a particular printed circuit assembly. The tier-one

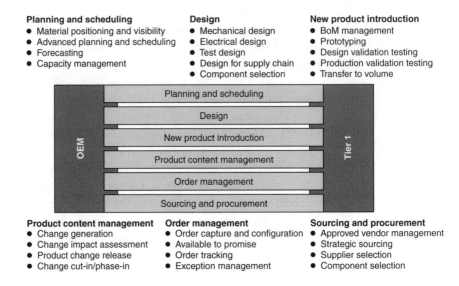

Planning and scheduling
● Material positioning and visibility
● Advanced planning and scheduling
● Forecasting
● Capacity management

Design
● Mechanical design
● Electrical design
● Test design
● Design for supply chain
● Component selection

New product introduction
● BoM management
● Prototyping
● Design validation testing
● Production validation testing
● Transfer to volume

Product content management
● Change generation
● Change impact assessment
● Product change release
● Change cut-in/phase-in

Order management
● Order capture and configuration
● Available to promise
● Order tracking
● Exception management

Sourcing and procurement
● Approved vendor management
● Strategic sourcing
● Supplier selection
● Component selection

Figure 5.1 Critical collaborative processes.

supplier should be able to communicate back to the OEM the supplier's capacity and the availability of material, so the OEM knows whether or not the supplier can meet the order requirements. Also, when an OEM changes an existing product or sub-assembly, the tier-one supplier should be able to report back that it has implemented the change and can produce the revised product.

The second step toward realizing all the benefits of collaboration is to use two-way communication to support joint decision-making. For example, when the tier-one supplier communicates that it is unable to fulfil an order, the partners work together to determine when the order can realistically be fulfilled – to their mutual benefit – by considering other priorities and available material, for example. Or when an OEM changes a product component, the partners work together to determine when the change should be phased in, taking into account demand for the product and inventory goals. Six critical collaborative processes have been identified (see Figure 5.1).

It can be difficult to establish collaborative processes because the supply chain is so dynamic. Demand is extremely difficult to forecast. Product and component life cycles are shortening. Product changes come fast and furiously from both demand and supply side. Manual processes cannot keep pace. To achieve collaborative status in this environment, supply chain partners need new technology in the form of supply chain software solutions.

While many of the collaborative capabilities identified in Figure 5.1 are available, not all exist today for fully built-out, tried, tested and standard products. The rapid advance in Internet-enabled capabilities, however, means that full collaborative capabilities are not far away.

As the need for collaboration becomes more important, OEMs and tier-one suppliers implement the limited capabilities currently available. Those that wait for fully built-out solutions will be at a serious competitive disadvantage to those that are already piloting capabilities and working with software companies to drive the development process. Early adopters of new supply chain software will reap the following benefits:

● reduced time-to-market and volume
● improved delivery performance
● increased supply chain efficiency
● reduced inventory excesses and obsolescence
● increased flexibility and customer service

Beyond collaboration between an OEM and its contract manufacturer, synchronization extends the collaborative processes up and down the supply chain to multiple customers and suppliers. An OEM, therefore, would communicate product availability and supply plans or product content changes up its supply chain to distributors and the rest of the channel. At the same time the OEM would communicate its demand requirements – based on an upstream forecast – or product changes down to its tier-one suppliers, and so on through the supply chain. In this fashion each member of the supply chain can see the same accurate and up-to-date data, enabling joint decisions to be made that can optimize the performance of the entire supply chain. Of course, for supply chain synchronization to be successful, the supply chain as a whole must work toward common goals.

While these relationships are up and down the supply chain, there is an added dimension – OEMs typically use the services of 2–20 tier-one suppliers, who have their own tier-one suppliers, creating even more complexity. This complexity, while creating challenges, also creates opportunities. As individual partners develop interface solutions, they can be replicated across all other relationships. We call the opportunity for synchronization across multiple supply chains the 'network effect'.

The network effect builds upon the benefits of synchronization through the use of standard tool sets that can drive process consistency among all supply chain partners. When every partner uses the same software across multiple relationships, with standard information formats and processes, most inefficiencies can be eliminated. Supply chain planning software by i2 and product content collaboration software by Agile Software are examples of such software. And even when different partners use different software, common standards for data and information structures and exchanges can drive efficiencies, for example, RosettaNet in the electronics industry.

Naturally, the degree to which companies choose to standardize or differentiate their processes is a strategic decision. While consistent tools and processes across the supply chain can create value, some companies may have a strong desire to differentiate themselves to drive value over and above brand management, customer management, and product innovation.

As the Network Effect extends across an industry, significant value can be extracted from the supply chain because tier-one and other suppliers will not have to operate multiple systems, solutions, and processes to support their different OEM customers. Most implementations today, however, are beginning with only a few supply chain partners. But as others see the benefits, this structure will be more widely adopted. We believe that collaboration will be a significant factor in differentiating competitive advantages over the next few years, as industries continue to focus on speed-to-value and the collaborative capabilities of supply chain software continue to develop.

Products like Agile Anywhere, which take advantage of the network effect, also drive its proliferation.

Conclusion

In the past, using traditional product data management systems and exchanging engineering data with suppliers during the design process was difficult, slow and geographically limited. Managing the design

CASE FILE

Agile Anywhere is rapidly becoming the standard solution for product content management in the electronics industry. Agile Software, founded in 1995, has established a customer base in excess of 600 OEMs and tier-one suppliers, including Solectron, Flextronics, SCI Systems and Jabil Circuit, in the electronics and high-tech industry.

Agile Anywhere manages product content information, bills of material, drawings and AVLs, beginning at the point at which the product structure is created, that is, after the design process has been initially completed. The collaborative nature of the product enables OEMs and tier-one suppliers to speed products into volume production by retaining central control over product information while providing access to that information across the supply chain; by providing a tightly integrated change management process; and by automating the process of updating product information in the local ERP systems.

process across the supply chain posed challenges to both profitability and customer service. Common scenarios included product quality problems, customer specification changes, process capability issues, and supplier substitutions and material shortages. These challenges often resulted in late product introductions, complex approval processes, distraction of high-value skills, systems compatibility issues, challenges in data sharing, coordination issues between geographically dispersed teams and higher costs.

Today's Internet technologies require less product data manipulation, carry greater data integrity, reach a broader global design community and can be implemented in shorter time frames. They allow participants both within and across organizations to collaborate during the design process, and to reach out to customers for feedback. Manufacturing companies that adopt new technologies will find their competitive position strengthened by bringing more suitable products to market faster and at lower cost.

Just as the eEconomy is revolutionizing product design, so too is it changing the way products are being manufactured. The manufacturing landscape is evolving from a series of discrete monolithic organizations into interconnected specialist organizations. The virtualization of the manufacturing function is driven by the combination of seamless integration of business systems, enabled by Internet technology, and management's focus on collaborative integration. For the many products that have become virtual, such as Internet distributed music, digitized images, and written publications, information technology has been substituted for the physical assets in the supply chain. But for manufacturers of physical products, there are new decisions to be made. With Internet-enabled real-time visibility and control of manufacturing decisions, the ownership of assets for some is becoming less important. While more and more companies are choosing to outsource their manufacturing to leverage the core competency and economies of scale of contract manufacturers, those companies that choose to retain their manufacturing capabilities are discovering new opportunities to capitalize on the eCommerce connectivity.

There can be no disputing the Internet's extraordinary properties that provide major performance improvement opportunities for many supply chain activities. However, it is also true that the physical nature of the majority of products means that exceptional supply chain management has an equally crucial role to play in the success of any eCommerce venture. The Internet offers the perfect communication channel to facilitate the superior coordination that is characteristic of the most exceptional supply chains.

6 Learning to synchronize supply chains through eMarketplaces

Introduction

There has been a dramatic growth in eMarketplaces, but we are not convinced that the dynamics that drove this phenomenon are sustainable. It is also not certain that the initial reasons for setting up eMarketplaces reflected where their true value would ultimately be. The initial reason for establishing marketplaces has shifted from extending procurement to a much broader range of opportunities for supply chain collaboration and synchronization. However, the opportunities are broadly misunderstood. While not simple or easy to implement, eMarketplaces can act as a fast-track mechanism for creating much greater integration, collaboration and synchronization between organizations.

Most eMarketplaces were set up by individual companies, industry invaders and industry consortia. The original eMarketplace agenda has evolved considerably since they were first proposed. Clearly, this trend has created more marketplaces than industries or business opportunities. The number of marketplaces is expected to grow to over 1000 realistic ventures before consolidating to fewer than 300. Many companies have scrambled to become part of these new marketplaces and a considerable amount of manoeuvring for position has occurred between multinationals, software vendors, systems integrators and investment banks. The main challenge marketplaces face today is to build capability and create enough liquidity to justify scaling up to profitability. Despite claims of significant value creation opportunities from these marketplaces their main current characteristic is their ability to absorb costs. We expect to see significant successes in some eMarketplaces, but also significant failures in many more as the realities of managing complex relationships and transactions sets in.

In this chapter we explore key issues associated with eMarketplaces and future supply chain collaboration and synchronization:

- How have eMarketplaces evolved and what are the expected future waves of change?
- What will drive success in eMarketplaces?
- How can eMarketplaces create the path to supply chain collaboration and synchronization?
- What will an eSynchronized world of eMarketplace competition look like?
- What are the key challenges and lessons learned from eMarketplace led collaboration and synchronization?

eMarketplace evolution and future waves of change

Even in the relatively short lifetime of eMarketplaces, we have witnessed several distinct waves of change (see Figure 6.1). After the initial 'dotCom invasion' of multiple industries, traditional multinationals engaged in a backlash against the invaders. Companies are now seeking to adopt a new business philosophy, where they combine the best of the dotCom technology and the best of the assets of traditional multinationals to create new business models.

We are likely to see widespread failure in eMarketplaces that are unable to build sufficient liquidity and scale. Indeed, such is the expectation of breakdown, that research companies are now publishing lists of

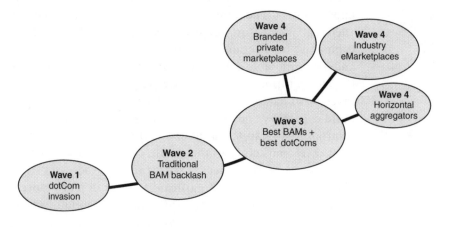

Figure 6.1 Four waves of eCommerce evolution – responses to eCommerce.

eMarketplaces that grade exchanges according to the likelihood of their failure. The impact of the multinational backlash against dotComs has been twofold. First, it had reduced the likelihood that dotComs will secure adequate funding by discouraging investors and delaying initial public offerings (IPOs). Second, it has extended the 'time-to-market' of industry eMarketplaces as multiple companies try to agree standards and relative positions.

dotCom invasions

The first wave involved dotCom start-ups entering marketplaces with bright ideas, sizable financial backing and superior technological 'smarts'. Typically, the invaders targeted industries that were seen to be slow moving and insular. In these industries, newcomers had a significant opportunity to use new value propositions to 'grab' market share from the traditional industry players. The incumbent organizations were generally caught off-guard and hence were slow to react. These dotCom invaders took on various forms:

- **Standalone** – generally standalone marketplaces were set up by a small number of people who saw promising business opportunities to create value. ChemConnect and CheMatch recognised the chance in chemicals, and Sciquest and Ventro saw the opportunity in laboratory equipment. Standalone companies generally had a good mix of industry and technology experts guiding their new business models.
- **Networked start-ups** – companies like Internet Capital Group (ICG) set up holding companies with strong venture capital funding to support the development of networked start-ups where there would be opportunities to share technology and infrastructure across the group. ICG, for instance, established a variety of industry vertical plays, such as eChemicals, Paper Exchange, and horizontal plays, such as VerticalNet and ICG Commerce.
- **Multinational spin-offs** – several major multinationals moved their people into new dotCom-style marketplaces, such as MetFabCity from Praxair. The intention was to capture significant market share ahead of competitors and to close out the attractiveness of start-ups to companies seeking to invade the industry.

The dotCom invaders were sufficiently effective that they attracted the attention of senior executives in incumbent companies. In some cases, these executives became convinced that there was very little they could do to compete against these invaders. In some other cases, it was the intervention of investment bankers that prompted whole industry consortia to compete against the invaders.

CASE FILE

The chemicals industry was a target for industry invaders. It was bombarded with several dotCom start-ups from an early stage in B2B eCommerce. The invaders included eChemicals, ChemConnnect, CheMatch and Chemdex. Initially, the traditional chemicals companies did not perceive the start-ups as a threat. In fact, they did not place a high priority on the adoption of eCommerce at all. Despite the participation of many of the BAMs in these industry exchanges, they were relatively slow to develop their own eCommerce strategies. Within only two years of implementation, billions of dollars worth of online transactions were going through these new dotCom companies. Chemdex, for example, in 1999, after only two years of operation, had signed up in excess of 130 suppliers, and had more than 300 000 products in its catalogue – five times that of the industry's biggest traditional catalogue. It was some time before the dotComs were challenged by the traditional chemical companies.

CASE FILE

The retail sector too was hit by the dotComs. Entrepreneurs and small retailers started conducting business online from as early as the late 1980s. Peapod was the first online grocery home shopping service in 1989, Amazon opened the virtual doors to the world's largest bookstore in 1995, eToys was founded in 1997, and Australia's IT eTailing was pioneered in 1997 with the online version of Harris Technology.

During 1999, there was a bombardment of entrepreneurial start-ups. The majority of leading traditional retailers, the retail BAMs, had either not yet established Web sites of their own, or if they had, their sites often lacked the presence and finesse of the pure dotComs' Web sites. This was so even to the end of 1999, which meant that many BAMs missed out on the skyrocketing of growth of online sales during the 1999 Christmas season. Conversely, Amazon.com recorded $16 million of sales in one day and shipped 20 million items during that period.

Traditional multinational backlash

In the second wave, the traditional industry players reacted against the invading dotComs, software vendors and other industry outsiders. As the BAMs recognized the potential value loss to the dotCom invaders they

CASE FILE

In the online trading arena, E*TRADE Securities Incorporated began to offer online investing services through America Online and CompuServe in 1992. The organization reaped early mover advantage and experienced exponential growth in demand for its online services with the launch of its own Web site, http://www.etrade.com/, in 1996.

started to question the power that dotComs and software vendors were commanding in their industries. The consensus among the incumbents was that the valuations that were open to dotComs were not justified by the value they added.

The traditional BAMs started to discuss alternative strategies for retaining the value in their industries, forming their own eMarketplace companies to rival the dotComs. Industry consortia were formed, such as Trade-Ranger in the oil industry, Elemica in chemicals, Covisint in automobile and Omnexus in plastics. The interesting question was whether these multinational consortia could hold together. It is always easy to unite partners out of fear of a third party, but it is harder to keep them together when that fear has passed.

CASE FILE

In the chemicals industry, Dow Chemical, Du Pont, BASF, (and IBM and Accenture) agreed to come together to ensure that they would dictate the direction of the industry rather than be dictated to by the dotComs. Initially, together with Bayer, and Ticona/Celeanese, they established a thermoplastics eMarketplace – Omnexus – to rival the existing plastics eMarketplace – Plasticsnet – that subsequently lost much of its IPO value and its trading volume. The major BAMs also started an industrial chemicals exchange – Elemica – taking on dotCom start-ups such as ChemConnect, eChemicals and CheMatch.

In the early days of 2000, eMarketplaces were being encouraged to expand by highly favourable stock market valuations. However, this optimism was short-lived and major stock market readjustments took place which hit the dotComs very hard as IPOs were cancelled and stock prices fell sharply.

CASE FILE

In the retail sector, industry players have announced several large exchanges, the two majors being WorldWide Retail Exchange (WWRE) and GlobalNetXchange (GNX).

Eleven major retailers formed a new Internet business-to-business (B2B) marketplace – WorldWide Retail Exchange. Its members now include 18 major retailers from around the world: American retailers Albertson's, Best Buy, Target, Kmart, Safeway, CVS, Gap, JC Penney and Walgreens; UK retailers Marks & Spencer, Tesco and Kingfisher; Jusco of Japan; Belgium's Delhaize; Royal Ahold in The Netherlands; Auchan and Casino in France; and Woolworths Limited in Australia. This collaborative partnership will enable transactions between retailers operating in the food, general merchandise and drugstore sectors. It is to be an open exchange that provides publicly available item data together with private price and promotion information between multiple buyers and sellers. The exchange will also provide auctioning services.

CASE FILE

Covisint was founded by Ford, General Motors (GM) and DaimlerChrysler and will be supplemented by the planned merging of Ford's AutoXchange, its joint venture with Oracle, and TradeXchange, the online procurement system that GM operates with Commerce One into Covisint. The annual purchasing value from these three companies is in excess of $240 billion, giving potential for massive savings, even if they are only of the order of a fraction of 1 per cent.

Renault/Nissan has subsequently joined Covisint, which is open to all auto manufacturers and suppliers, and since December 2000 has operated as a limited liability company.

The extensive financial support provided to dotCom start-ups withered away. Spectacular failures among the early dotComs resulted – the 'dotBoms'. The key driver of this phenomenon, quite apart from the backlash of Wave 2, was that the fundamental business propositions and models that made sense in the early world of dotCom hype simply did not have the foundations to sustain them once the 'bubble burst'. Many of the offerings were relatively narrow, and commanded realistic sized markets that were ultimately too small to justify the multi-million dollar capitalizations and valuations originally placed on them. An important lesson to be learned from the failure of many of the dotComs has been the need to recognize failure early and to be brave enough to close down failed eVentures fast.

CASE FILE

Among the many deaths of dotCom organizations, Boo.com, an online designer goods company, was one of the most spectacular. Boo.com built a strong brand image, had an impressive Web site and used leading edge technology. Unfortunately, the technology was too advanced for most home computers, which meant that most potential online customers could not place orders. As a result, its $220 million start-up capital was exhausted after only a couple of months of operation.

The best multinationals working with the best dotComs

In the third wave, the multinationals have recognized the weaknesses in most dotComs as well as their gaps in capability. The smarter multinationals are actively seeking to team with the best dotComs. They have learned that they must seek out suitable partnerships with some of the best dotComs and software vendors to create the capability or fill their own capability gaps. We are already witnessing leading organizations entering the third phase in their strategic response to eCommerce. This is the wave of joint ventures between dotComs and traditional multinationals, partnering to restructure their industries and establish standards and protocols so that they can run their industries through eMarketplaces – 'the best with the best'. Early examples of this trend were over 30 chemicals companies investing in ChemConnect. More recently several multinationals have announce joint ventures and alliances with dotComs and software vendors to create new capabilities, such as Eastman with GLog to form ShipChem.

The proliferation of exchanges and the different types of eMarketplaces is illustrated by Table 6.1. This table shows just a small part of the eMarketplaces that have been set up in the last few years. As we are seeing, there is a very big difference between setting up and delivering success from one of these new ventures.

New B2B eMarketplace models

Considerable debate is occurring over what comes next. Certainly some eMarketplaces will fail to reach sufficient size or become viable, but the fourth wave will also have its successes as new business models evolve and adapt to increasing opportunities. We believe that this wave will be characterized by a combination of different plays: industry vertical eMarketplaces, company mini-hubs and horizontal aggregators.

CASE FILE

In 1999, Wal-Mart, America's largest retailer with $160 billion sales in 1999 and number 2 on Fortune 500 list, built its own Web site. Its independent efforts to join the world of eTailing proved to be an expensive failure, with its Web site rated last for overall service by Resource Marketing of Columbus, Ohio. Among the site's many flaws, Walmart.com's search engine was inadequate and its responsiveness to consumer emails was poor. A measure of the site's inability to meet consumer expectations came when Walmart.com discouraged customers from placing Christmas orders after 13 December that year.

Despite Wal-Mart's pains to fix the site's problems for its re-launch on 1 January, problems persisted. Wal-Mart has since embarked on a new course into eTailing. It formed two strategic alliances to gain the expertise it needed to prosper in the electronic environment. An agreement with AOL helped create a co-branded low-cost Internet service, and a joint venture with Accel Partners, a Silicon Valley venture capital firm, provided essential Internet know-how and technological expertise.

In-store kiosks are to support AOL and Wal-Mart's Internet service in order to introduce shopping on the Web to potentially millions of customers. While the Wal-Mart brand pulls customers to the AOL portal, AOL provides a place where customers regularly visit and a cost-effective method for getting the business online. The joint venture with Accel, formed in January 2000, spun off Walmart.com as an independent online retailing operation, in which Wal-Mart owns at least an 80 per cent interest. Accel brings to the partnership its Internet expertise and Wal-Mart brings strengths such as purchasing power, a network of physical stores where the site can be promoted and goods can be returned, and most importantly, its name. Despite the early problems, Wal-Mart's reputation continues to pull traffic to the site even without promotion. According to Media Metrix, Walmart.com was the 46th most visited site in February 2000, attracting 1.4 million new visitors.

Kmart, America's third largest retail, has also formed a joint venture to capitalize on the opportunities offered by eTailing. Working with Yahoo! and Softbank Venture, Kmart has created a spin-off online company with its own brand name Bluelight.com.

Major Industry Vertical eMarketplaces
In the old economy, a typical multinational organization was made up of multiple functions and processes including planning, supply, demand, fulfillment and infrastructure, which were generally managed in-house. In some companies and industries we have started to see substantial organizational dismantling and cross-company reassembly of certain functions and processes, for example, BP outsourcing its finance and

CASE FILE

In the chemicals industry, a group of the leading traditional companies planned their domination of the industrial chemical market. The strategy adopted was to partner with an existing eBusiness to gain eCommerce expertise and reduce start up costs. ChemConnect was selected over eChemicals and CheMatch. Each of the leading traditionals took a 10 per cent equity share in ChemConnect and 5 per cent shares were offered to other players.

ChemConnect has now been established as the industry standard. The industrial chemicals and plastics marketplaces have aligned their standards with ChemConnect's and it is anticipated that an agricultural marketplace, an industrial gases marketplace and elastomer marketplace will also be established by the consortium. In essence, the chemicals industry is expected to be divided into six marketplaces that will be controlled by the key industry players in ChemConnect. While it is still not clear how this marketplace is emerging, key challenges for ChemConnect will be creating liquidity in the marketplace, building scale and achieving profitability.

accounting and IT. We expect this trend to be reinforced as we see the formation of large vertical industry eMarketplaces across a range of industries. Different industries have taken different routes. Sometimes we are seeing one major vertical structure per industry, as with Trade-Ranger in the oil industry. In others, such as food and retail, we are seeing competition between groups of large multinationals to be the dominant eMarketplace. In industries like chemicals, both big company eMarketplaces, such as Elemica, and smaller company eMarketplaces, such as Envera, have emerged. In some regions, particularly places like Australia, we are seeing the emergence of eMarketplaces based on geographic basins.

CASE FILE

Transora is the largest vertical exchange in the food and consumer packaged goods industry and is leading the transformation of the consumer products and services industry through a global, open, standards-based, industry-led marketplace that delivers breakthrough value for all industry participants. It provides end-to-end supply chain services globally as well as connecting buyers and suppliers to other retail exchanges currently under construction. Transora's founders are over 50 of the world's largest food and consumer packaged goods companies, including Sara Lee, Unilever, Coca Cola, Gillette, Hershey Foods, Nestlé, General Mills and Procter & Gamble. Founded in March 2000, Transora uses i2 and Ariba technologies.

Table 6.1 eMarketplaces.

Industry Consortia	Independent Trade Exchanges		Private Exchanges
● Bond Book	● Allplastix.com	● L-Trans	● Alstom
● Covisint (auto exchange)	● Antenna network	● LumberLogix	● Arbed
● CC Markets (chemicals, Germany)	● appliancezone	● Metalsite	● Banamex
● Computer Exchange	● basins.com	● MetFabCity	● Barclays Bank
● CITE (construction – ANZ)	● biospace.com	● NutaBud	● Concert
● CPGMarket.com (Europe)	● Buyproduce.com	● Nymex	● Conoco
● Defense Logistics Agency	● Cephren	● omnicell.com	● DaimlerChrysler dealer portal
● Exostar (A&D)	● ChemConnect	● Onbedrock.com	● Dell Computer
● ForestExpress	● ChemCross	● PaperExchange	● Electronic Risk Exchange (ELRiX)
● EnergyLeader.com	● ChemRound	● Personic	● eSource (CIBC)
● Hospitality Exchange (Europe)	● Covation	● Portum	● finlay
● IES (food and packaging goods – ANZ)	● Dairy exchange	● PrintCafe	● Global Insurance Broker
● Industrial Vehicle Exchange (I.V.Ex)	● eBay	● Proventia	● Hong Kong Telecom
● Logestica (S.A. utilities)	● ebizmix.com	● Quaris	● ISO New England
● Mercado Electronico	● eBricks.com	● rooster.com	● London Stock Exchange
● Metique	● eDownstream	● SEALS	● NorthCarolina@Your Service
● Quadrem (metals and mining)	● eExchange	● Security First Technologies	● Oil & Gas Marketplace
● Omnexus (chemicals)	● Engineroar	● Shoplink	● Paccar
● Rail Exchange	● ePlas	● *simply*engineering	● Partek
● Steel Exchange	● epylon.com	● SmartBearing	● Sharp Electronics
● Telco Matrix	● FuelQuest	● Via	● Swisscom
● Trade-Ranger (energy)	● HoustonStreet.com	● Web eSTP	● Thomson
● Transora (food and packaging goods)	● ICG Commerce	● WebModal.com	● via World Network (AC)
	● infrastructureworld	● Whitegoods Appliance	● Whirlpool
	● i2i	● yet2.com	
	● iSuppli		
	● JusticeLink		

CASE FILE

In September 2000, a consortium of the leading freight railroads in North America, made up of Burlington Northern and Santa Fe Railway, Canadian National Railway, Canadian Pacific Railway, Norfolk Southern Railway and Union Pacific Railroad founded Rail Marketplace. iRail.com and GE Global eXchange Services (GXS) are also part of the alliance, whose goal is to establish the leading electronic marketplace offering supply chain optimization services for the rail and rail-related transportation industry. Using GE GXS internally developed capabilities, Clarus and other related technologies, this Internet service aims to enhance value for all participants in the supply chain through more efficient and effective procurement and sourcing of materials, with services to be added over time.

CASE FILE

Omnexus is the neutral independent eMarketplace for the plastics industry. Initially designed and built to meet the demands of the highly competitive injection moulding market, it is a seller-supported and buyer-designed marketplace with offerings that include resins, machinery and tooling and related supplies and services. Omnexus was founded in June 2000 by Dow, DuPont, BASF, Bayer and Ticona.

CASE FILE

Quadrem is an independent, neutral and open procurement marketplace for the mining, minerals and metals industry. It will allow buyers and sellers to trade all goods and services. Value added services such as logistics, finance and payments will also be offered. Quadrem was founded by Alcan, Alcoa, Anglo American, Barrick Gold, BHP, CODELCO, CVRD, De Beers, Imerys, Inco, Newmont, Noranda, Pechiney, Phelps Dodge Corporation, Rio Tinto and WMC. The shareholder agreement was signed in October 2000.

CASE FILE

Trade-Ranger is a global Internet marketplace dedicated to buying and selling materials and services used across the supply chain of the energy industry, including the upstream, downstream, retail and petrochemical sectors. The exchange was founded in May 2000, uses Commerce One and i2 Technologies and has $125 billion in committed spend. The industry participants of this independent marketplace include Royal Dutch Shell, BP, TotalFinaElf, Dow Chemical, Conoco, Motiva Enterprises, Phillips Petroleum, Equilon Enterprises, Mitsubishi Corporation, Repsol YPF, Statoil, Unocal and Tosco.

The vast majority of industry exchanges have been fuelled by eProcurement software vendors and therefore conduct some type of procurement role. However, it is increasingly clear that it is not going to be possible for vertical industry eMarketplaces to aggregate procurement. The question then becomes: 'What can they do to add value and extract returns within industry?'.

Two viable options exist. First, we expect to see industry eMarketplaces taking on utility roles in that they help to connect up an industry. With benefits in the form of transaction and interaction efficiencies, industries will have common standards and unambiguous communications. Collaboration and box-to-box transactions will be made easier and more effective.

Second, we expect to see industry eMarketplaces taking on industry-level supply chain optimization and 'demand chain synchronization' activities. They will look for opportunities, from a market perspective, to optimize logistics operations and infrastructure. This would entail the sharing of forecast and capacity information and industry level management of assets.

The success of these industry eMarketplaces will hinge on the participating organizations' abilities to operate new business models, manage the relationships between themselves, and manage the complexities of their industries.

Branded Private Marketplaces – Big Company eHubs
The larger multinationals that feel strong enough will set up eMarketplaces for themselves and their customers and suppliers. These companies will want to offer aggregation capabilities to immediate customers and suppliers in their value chains. They have a strong opportunity to offer their small and medium size customers and suppliers, access to the leverage that they themselves command through industry eMarketplaces. This includes access to procurement, fulfillment and technical support networks. For example, a large oil and gas company could help its distributor customers to buy transport at better rates through an industry eMarketplace than they could themselves negotiate.

Several other drivers potentially exist for the emergence of company mini-hubs. For example, an organization may contend that its geographic positioning warrants that it stand apart from its fellow industry players. This may occur if the organization, and its key customers and suppliers, are to some degree physically isolated from the business centre for their industry, or if there exists significant physical barriers between them.

The organizations that have been most active in setting up company mini-hubs, like GE and IBM, seem to combine two characteristics. Firstly, they are generally big enough to do this and secondly, they tend to be sceptical about the ability of industry consortia to hold together relationships and build eMarketplaces on a realistic time-scale. We believe that companies will find this approach to eMarketplaces increasingly attractive in the future.

CASE FILE

GE Global eXchange Services (GXS) operates one of the world's largest B2B eCommerce networks. It has in excess of 100 000 trading partners and handles 1 billion transactions annually, representing a total of $1 trillion in goods and services. The company counts 60 per cent of Fortune 500 companies among its customers and has a presence in 58 countries. Key customers include Coca Cola, DaimlerChrysler, Eastman Kodak, JC Penney and Lockheed Martin. GE GXS has engineering teams in the USA, Ireland, the United Kingdom, India and the Philippines. GE Global eXchange Services is part of GE Information Services, Incorporated, a wholly owned subsidiary of the General Electric Company.

GE GXS helps businesses drive cost from their supply chains with three portfolios of electronic commerce products and services:

- GE Integration Solutions – provides software that permits secure and reliable communications between different business applications.
- GE Interchange Solutions – improve quality and efficiency in a supply chain by automating paper, fax, telephone and email transactions.
- GE Marketplace Solutions – business applications and technology infrastructure for the development, integration and service of high-volume B2B eMarketplaces.

Horizontal eMarketplaces
Horizontal aggregation is likely to be the most common play in Wave 4, due primarily to the competition regulations affecting vertical aggregation. Other drivers include:

- Geographic circumstances: organizations located in geographically isolated regions may find that the only avenues offering the potential to achieve critical mass are horizontal plays.
- Inter-industry activity: for every demand cycle in one industry, further down the 'food chain' there is the supply side of another industry – circumstances that offer great potential for cross-industry marketplace activity.

Horizontal procurement aggregation plays will emerge such that the market share in any individual industry will be kept down so as to avoid the regulators. Organizations will also look for cross-industry optimization opportunities in areas such as education, training and asset management. Perhaps the most obvious example is fulfillment. From a logistics point of view, there appears to be enormous opportunity and incentive, for say 3PLs and 4PLs, to start building services to create super fulfillment hubs that service multiple, possibly interconnected industries.

CASE FILE

Metiom (previously Intelisys) was founded in 1996 with investors such as Forstmann Little & Co and the Chase Manhattan Bank. It trades numerous services such as accounting, advertising and legal across a number of industries. Metiom differentiates itself by focusing on small to medium enterprises. By early 2000, the exchange was trading in the vicinity of $2.5 million per day from around 600 suppliers.

CASE FILE

The merger between Accenture's eProcurement venture, ePValue, and ICG Commerce instantly created the largest, and first truly global, comprehensive eProcurement services provider in the marketplace. The new entity operates under the ICG Commerce (ICGC) name. The agreement incorporates key operating and investment commitments by ICG Commerce, Accenture and Sun Microsystems, a strategic partner in, and an equity owner of, ePValue. The combination of ePValue's multinational operations in North America and Europe, with ICG Commerce's operations and breakthrough comprehensive end-to-end eProcurement approach, has over $10 billion in actual purchases. ICGC aims to deliver the most effective procurement capability available through superior speed, reach and scale. It provides large multinationals and mid-sized companies with immediate access to fully operational, turnkey global eProcurement capabilities. ePValue brings a total category management outsourcing capability together with ICG Commerce's exchange-based offering that requires no up-front capital investment in software. ICGC employs professionals whose exclusive focus is procurement, including specialists in strategic sourcing, auctions and Internet-based procurement, and category management. The operation is supported by a global customer care organization. The merged customer bases nearly doubled the purchasing power of ICG Commerce.

CASE FILE

CorProcure is an Australia-based horizontal exchange which was announced in July 2000. Some of Australia's largest companies have collaborated to establish CorProcure, including Amcor, AMP, ANZ, BHP, Coca-Cola Amatil, Coles Myer, Foster's Brewing Group, Goodman Fielder, Qantas and Telstra. Participants will include Australia Post, Orica, Pacific Dunlop and Wesfarmers. It is estimated that AU$8 billion of transactions will go through the exchange over the next two years. The products and services that will go through the exchange include: advertising, computer services, energy, human resources, facilities management, fuel, legal services, insurance, office and cleaning supplies, and telecommunications.

CASE FILE

Making international shipping easier for small and medium-sized businesses, FedEx Global Trade Manager is an Internet service used by US-based customers who arrange shipments to or from the USA, Canada, the United Kingdom, Hong Kong and Puerto Rico. Recognizing the difficulties of filling international orders given the complexities of shipping across borders, this online service helps shippers identify and prepare appropriate import and export forms based on commodity descriptions and countries involved, highlighting potential customs issues before they become a problem. Global Trade Manager is the latest addition to the portfolio of eCommerce solutions offered by FedEx, following the recent introduction of FedEx eCommerce Builder, another online service aimed at helping small to medium-sized businesses manage online stores.

What is being played out is more a knowledge and information game than a game of assets. Clearly access to assets is required, but the very smart organizations will come in and try to optimize flows within industries, and between industries. Perhaps the biggest challenge will be to manage the horizontal plays in a market unit-driven environment.

What will drive success in eMarketplaces?

The expected role of eMarketplaces has changed significantly over time. Initially they grew out of the thinking of eProcurement vendors such as Ariba and Commerce One. Many started these new ventures by conducting auctions to capture value quickly, such as the Ford Tyre auction. Now the emphasis has started to change, with a much greater

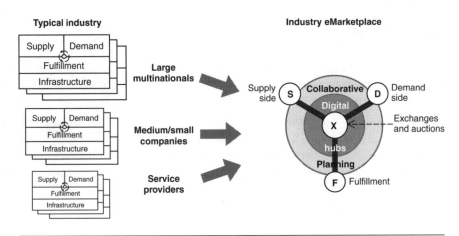

Figure 6.2 Disaggregation and reaggregation of demand, supply and fulfillment transactions.

expectation that they will provide industry-level services and standards covering a wide range of supply chain and infrastructural capabilities. It is quite useful to think about these marketplaces as financial exchanges applied to other products and services. We see the same level of disaggregation and reaggregation of supply and demand as one would expect to see in a financial market (see Figure 6.2). The big difference between these eMarketplaces and financial markets is clearly the need to include fulfillment costs and services into the overall optimization equation, for example, you do not buy a product unless you know how much it will cost to ship it to you.

Sustainable business models and realistic stakeholder expectations

eMarketplaces will need a sustainable business model based on offerings that will realistically command significant market shares and incentives for users to place a wide range of transactions through them. One of the surprises early in 2000 was the initial naivety of many of the revenue and profitability models that were promoted as the business cases for some eMarketplaces. Quite a few marketplaces sought to justify their future on transaction revenue models that were entirely unrealistic. Fortunately, this situation started to change very quickly by mid-2000. Most eMarketplaces now understand they need a business model and business case that is robust and recognizes that sponsors will only fund the initial build and are unlikely to subsidize transaction fees in the medium term. The best eMarketplaces are recognizing that long-term viability is standalone and the time to create a sustainable business model is now.

Stakeholder expectations need to be clearly defined and managed. In our experience, the most likely outcome of multi-partner ventures is one of increasing costs, declining benefits and reducing functionality relative to initial expectations. Stakeholders must be aware of the realistic business plan and must have total buy-in to the business model, the partnerships, the commission structures and the value propositions on offer. A common consequence of over-optimism comes when reality dawns and the often 'all-important' second round funding is suddenly not forthcoming.

The high levels of eMarketplace complexity, the volatility of the competitive environment and the critical nature of relationships in eMarketplaces raises the importance of having the right management team in position. Management must know their target industries and markets appropriately. They must understand the competitive dynamics and drivers such that they are able to lead and manage change effectively. Management must be fast to act and react, and be prepared to quickly abandon strategies, approaches or partnerships that do not deliver on expectations. They must establish a performance management system that enables timely responses. There is no right formula for selecting the right management team, except that you will know how it feels when you have not got it right!

Rapid building of key capabilities

Successful eMarketplaces must rapidly build a full range of capabilities to become the 'one stop shop' for an extended range of transactional and value-added services. It is not likely to be sufficient for an eMarketplace to sustain a niche set of offerings. eMarketplaces will need to balance the creation of liquidity in standard services with the need to continue the building of a more complete range of specialized services to capture market share and achieve the transactional volumes necessary to build longer term profitability and industry as well as customer acceptance (see Figure 6.3). In the early days of eMarketplaces, few companies had the vision to see what this breadth of services would consist of, indeed most were entirely focused on winning the eProcurement 'PowerPoint war'.

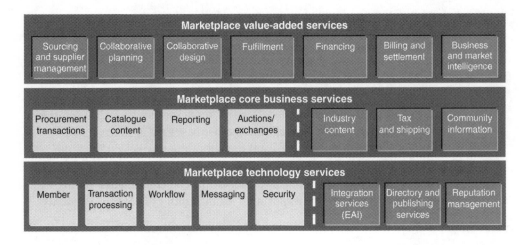

Figure 6.3 Footprint of eMarketplace services.

The realization of the need for a broader range of services grew strongly in mid-2000 and has led to a range of alliances and acquisitions, such as Ariba and i2, SAP and Commerce One. The software vendors have been very active in developing their strategies for filling gaps in their product ranges (see Figure 6.4), and there has also been a strong drive to understand how the key elements of eMarketplaces can work together. This is still very much 'work in progress'.

The real impact of this change in thinking has been a strong intent to combine the key supply chain elements of planning, procurement, manufacturing, design and fulfillment under common standards that unite an overall value chain. This is an ambitious vision that would have seemed inconceivable even two years ago.

Integration of activities with major stakeholders

It is one thing to set up an eMarketplace, but quite another thing to get participants (even with equity stakes) to participate. Indeed, many of the early eMarketplaces now seem naive about the challenge of 'onboarding' equity participants. The level of integration required between eMarketplaces and participating companies is potentially huge, but most eMarketplaces, and the companies investing in them, do not yet fully understand this.

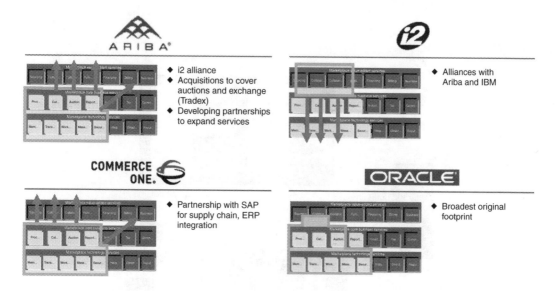

Figure 6.4 Technology providers building capabilities.

Long-term success of an eMarketplace will hinge more and more on the degree of integration between the eMarketplace and an ever-increasing range of partners and stakeholders. Seamless links must be established both upstream and downstream in the supply chain as well as with providers of specialized services such as financials, logistics services and possibly other marketplaces (see Figure 6.5). There is renewed emphasis on connectivity with both supplier and buyer ERP systems. The required levels of integration and open communication intensify the need for all stakeholders to have a common language and shared standards and protocols.

The integration challenge has many interesting implications for the different levels of participants in eMarketplaces. For the eMarketplace owners and software suppliers, the challenge is one of setting standards, building capability and encouraging usage. For participants the challenge is one of 'onboarding' or getting companies to build the interfaces they will need with their own systems to be able to participate at scale. For eMarketplace vendors, such as software suppliers and providers of specialist services, such as logistics and payments service providers, the challenge will be filling product and service gaps fast enough to maintain credibility with eMarketplace participants.

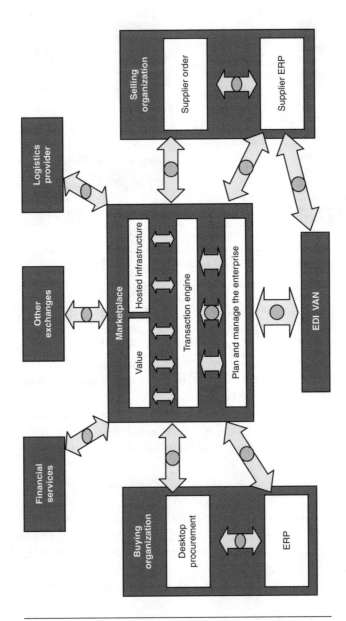

Figure 6.5 eMarketplace stakeholder integration.

Careful management of regulatory relationships

Antitrust authorities in the European Union (EU) and the USA are still struggling with the questions of how, and to what extent, business-to-business (B2B) eMarketplaces should be regulated. While they do not wish to obstruct progress in this area through the enforcement of overly restrictive regulations, there is the concern that competition could be significantly hampered if new eMarketplaces are allowed to develop without addressing potential antitrust issues.

CASE FILE

In August 2000, the European Commission ruled in favour of MyAircraft after its investigation into the eMarketplace's potential for price fixing. MyAircraft is a consortium of USA companies, including Honeywell, United Technologies and i2 Technologies, which provides an eMarketplace for aerospace after-products and services. The Commission reasoned that the eMarketplace faced strong competition from a relatively high number of other B2B marketplaces in the same sector, such as Exostar, the Boeing, BAE Systems, Lockheed Martin and Raytheon exchange group, and that third parties considered MyAircraft to be just one of several business entities through which companies transact business.

After its ruling in the case of MyAircraft, the European Commission was quick to emphasize that the case had not established a precedent for other B2B marketplaces. Other exchanges should therefore expect to be examined on a case by case basis. While MyAircraft was investigated under the European Union's Merger Regulations because it is a full-function venture jointly controlled by its parents, B2B marketplaces established by single companies would not fall under the regulation because there would be no concentration. However, the B2B marketplaces that fall outside the merger regulations may come under general EU Treaty rules on restrictive business practices, in which case there is no obligation to obtain prior regulatory clearance.

The European Commission's investigation has no doubt brought B2B exchanges into the spotlight, with the prevailing concern that if companies entering online exchange platform deals exchange price-sensitive or other sensitive information, they are effectively acting as a cartel. In the same way as they are unable to do so offline, big businesses should not be able to set up cartels online.

There is still much to be resolved around regulatory responses to eMarketplaces. In many ways, regulators have been slow to understand the dynamics of eMarketplaces and to decide on what is and is not acceptable. Many recent judgements seem to be more like holding positions rather than legal direction. However, this is understandable giving the loose nature of the definition around what individual eMarketplaces are trying to achieve.

In the absence of a definitive 'safe' model for B2B eMarketplaces, stakeholders must establish and carefully manage their relationships with the regulatory bodies from whom they require approval. Owners and operators should take all precautions to ensure that their marketplace is seen to be open and cooperative, and that there are sufficient measures in place to prevent any occurrence of anti-competitive practices.

The 'eMarketplace in a box'

With increasingly complex demands on eMarketplace capability, there is an emerging demand for 'eMarketplace in a box' solution offerings. 'eMarketplace in a box' offerings would essentially be a skeleton marketplace structure: a set of generic, partially configured components, ready to be populated and tailored to meet the specific requirements of the operators (see Figure 6.6). These turnkey solutions will serve to minimize the time and effort spent on integration by using a pre-selected suite of best-of-breed applications.

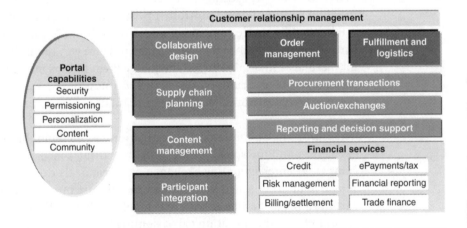

Figure 6.6 'eMarketplace in a box' solutions.

Such 'Marketplace in a box' solutions may also have a number of beneficial solutions:

- Share development costs for eMarketplaces across multiple industries/ horizontals.
- Define standards between industries where common eMarketplace solutions are used.
- Simplify the building of 'peripheral' activities in eMarketplaces such as logistics and payment services.
- Remove problems associated with custom build solutions in multiple eMarketplaces.

We expect to see a significant increase in demand for 'eMarketplace in a box' solutions as companies become increasingly frustrated due to the failure to deliver of early start-ups and industry consortia.

Synchronizing the supply chain through eMarketplaces

An unexpected but predictable side to eMarketplaces is the opportunity to accelerate progress towards an eSynchronized supply chain world. We have been developing synchronization solutions using the latest applications for several years. However, there has been little take-up due to lack of understanding or perceived relevance outside of high-tech industries. Moving forward, eMarketplaces look like being the next step towards eSynchronization. They are increasingly setting out an agenda that combines eProcurement with eDesign, eManufacturing, ePlanning and eFulfillment across a range of industries.

The requirements for increasing collaboration and synchronization are reasonably well understood. Major areas of work are being conducted on building supply chain planning and enterprise applications integration (EAI) solutions to enable individual companies and eMarketplaces to take the next step towards eSynchronization.

A key factor in this shift in the role of supply chain planning, EAI and eMarketplaces is the role of enterprise resource planning (ERP) systems such as those offered by Oracle and SAP. ERP systems will continue at least in the immediate future to be the backbone of both company and eMarketplace transaction systems. The key question is whether they should be high or low level in their functionality. Some companies are adopting a 'wall to wall ERP' strategy, choosing one vendor for everything. Others are building 'ERP Lite' solutions which have limited functionality beyond basic transactions and integration.

eMarketplaces represent a key step on a journey to supply chain synchronization but one that is difficult to deliver. We believe that the stage is now set for the best companies to exploit these collaboration and synchronization technologies as part of their strategy to dominate future transactions between individual companies and public and private eMarketplaces.

The role of collaborative supply chain planning solutions

Supply chain planning (SCP) incorporates a number of components, including forecasting, distribution resources planning (DRP), manufacturing resources planning (MRP2), transportation scheduling and manufacturing scheduling (see Figure 6.7). Initially most ERP systems had some of this functionality, but in many cases it was limited. Planning was therefore disjointed and piecemeal as ERP systems could only plan those parts of the supply chain with transactions on that particular system.

In recent years, SCP solutions have significantly improved in terms of both functionality and usability. SCP solutions now sit above the ERP transaction systems, lifting data such as orders, inventories and production quantities. The information is used to create plans for the entire supply chain and can be fed back down into the ERP systems for execution: a major step forward for supply chain synchronization.

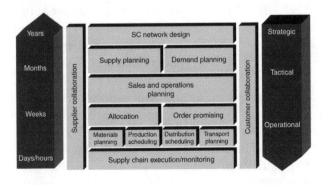

Figure 6.7 Components of supply chain planning.

In today's leading solutions, DRP and MRP2 have been merged into a single supply chain planning module. Demand planning (DP) tools have been developed so that it is now possible to build sophisticated event-driven forecasts that incorporate causal factors, track forecast consumption and monitor forecast accuracy. SCP and DP have been closely integrated so that the impact of a change of forecast on manufacturing and distribution can be seen immediately, rather than at a particular point in the sequential planning processes, as was the case in the past. This timely visibility enables scenarios to be tested and supply chain trade-offs analysed more rapidly than before. Exception-based planning, using exception alerting tools to plan, forecast or schedule, is also built in to all leading SCP applications. It is even possible to optimize the supply chain not just on the basis of cost but also on profitability and other key drivers.

The two dominant SCP vendors are i2 Technologies and Manugistics. Both provide a comprehensive range of functionality covering all the areas described above. Manugistics founded the first integrated supply chain management suite in 1986. i2 followed closely in 1988 and its 1999 revenues of $571 million won it the market leader position. SAP and Oracle are rapidly developing their own SCP solutions, SAP Advanced Planning Optimizer (APO) and Oracle APS. Vendors focused primarily on process industries rather than discrete manufacturing include Aspentech and Logility, and other players include Numetrix (now acquired by JDE), Synquest and Adexa (formerly Paragon).

Leading eMarketplaces will play a key role in the delivery of collaborative supply chain solutions. Truly collaborative supply chain planning cuts across business unit, company and even industry boundaries, and requires more than the simple passing of limited information back and forth between the parties. The high levels of information sharing and openness required for true collaboration can best be met through the use of seamlessly integrated systems and shared databases or information repositories. The Internet, and in particular eMarketplaces, are well placed to assume the pivotal role of information repository host, coordinator, and collaboration facilitator.

eMarketplace operators have the opportunity to provide users with access to a full range of supply chain planning applications and collaboration tools. It will no longer be necessary for individual organizations to invest in and maintain their own applications. Instead, the eMarketplace would provide the network services and the hardware and software infrastructure while applications service providers would provide applications and content.

Faster and more accurate communications, including the alleviation of many of the issues of version control and inconsistent standards, will result from utilizing the common technical platform and applications on offer through eMarketplaces. In turn, improved forecast accuracy and more effective decision making will enable significant reductions in inventory holdings and a higher degree of alignment with actual demand throughout the supply chain. Other components of the value proposition include the reduction in information technology costs and the inclusion of business flexibility into the information technology structure.

CASE FILE

i2 Technologies' TradeMatrix is a dynamic Internet marketplace that provides a one-stop destination for online collaboration and dynamic trading, electronic procurement, spot buying, selling, order fulfillment, logistics services and product design services. TradeMatrix provides an open digital community where customers, partners, suppliers and service providers gather to conduct business in real time, enabling companies to make more profitable decisions.

The supply chain planning arena is set for rapid expansion, particularly as eMarketplaces recognize its value in their offerings. Web-based, or eMarketplace collaborative planning will allow supply chain participants to create a 'virtual' store of inventory that each participant can access to satisfy customer needs from any available source. Full knowledge of availability across the supply chain will allow these participants to reduce costs through lower stocks and more efficient shipment planning. By allowing participants to operate supply chains at 'eSpeed' through the sharing of production, inventory, product and shipment status, companies can gain competitive advantage by beating others to new customers and markets. This type of development is crushing old paradigms of 'ownership' of key strategic, planning and operational information. The significant benefits of sharing information with business partners are enticing companies to build trust levels and release key information from their direct control.

CASE FILE

Sun Microsystems is taking advantage of the collaborative nature of the Internet by developing Web-based collaborative planning tools as a way to strengthen strategic relationships with key customers. These tools allow Sun to exchange forecast and product status information with customers on orders, shipments and promotions, and help Sun to manage their products through the entire life cycle. This capability has resulted in substantial reductions in lead times and forecast availability, improved inventory turns, increased customer satisfaction, and more efficient supply chain operations.

Linking ERP and EAI to eMarketplaces

A major barrier to synchronization within organizations has been the difficulty of sharing data and managing business processes across multiple IT applications, which are often running on different platforms in several locations. This problem is amplified when synchronization is attempted across a number of organizations. The current trend towards 'best of breed' application solutions means that this issue is not going away. Three or four applications within a single architecture are not uncommon, even before integration with legacy systems is considered. There is an urgent need for solutions that reduce the cost of data integration and allow business processes and workflows to be managed across application and organizational boundaries. A range of solutions are rapidly emerging under the collective banner of enterprise applications integration (EAI).

Accenture has defined EAI as: a set of technologies that enables the integration of end-to-end business processes and data (information) across disparate applications to increase the organization's and supply chain's ability to respond and adapt to change.

EAI is not merely technology, nor is it simply the automation of business processes across otherwise independent systems. Rather, it is a set of complementary technologies, working together not only to integrate business processes in the existing environment, but also to allow an entity to evolve quickly and painlessly as its IT landscape changes. EAI is a strategic technology.

EAI solutions are more than middleware. Middleware solutions typically focus on integration at the 'data level'. While this provides the connections that allow applications to see each other's data, it does not necessarily mean that the data can be understood or used. EAI solutions, in addition to middleware, will typically offer:

● Packaged interfaces ('connectors' or 'adaptors') which reduce the development effort in getting information in and out of databases or applications.
● Transformation engines, which reduce the development effort in translating message formats and routing messages.
● Process management engines enabled by workflow capabilities, which reduce the development effort of coordinating the overall flow of information between applications and databases. This also enables management of business processes spanning multiple applications.

Industry analysts believe that the opportunity in this area is huge, but at the close of 2000, there was still no 'silver bullet' for solving the integration problem. Emerging vendors, such as SeeBeyond (formerly known as Software Technologies Corporation (STC)), Viewlocity, Vitria, webMethods Enterprise, Neon, TIBCO, CrossWorlds, Extricity and others, have jumped to the forefront of this market.

The focus of the EAI vendors varies significantly. Figure 6.8 shows the distribution of vendors over two criteria. The first, business process management versus data/messaging management, refers to the functional focus of the vendor, and the second, intra-enterprise versus inter-enterprise, refers to the scope of the application.

	Intra-enterprise (eAI)	Inter-enterprise (B2B,B2C)
Business process management	● Active ActiveWorks™ ● CrossWorlds UAA™	● Agile Anywhere™ ● Aspect eXplore 2000™ ● Datasweep Advantage™ ● Extricity Alliance™ ● i2 Rhythm Link™ ● Parametric Windchill™ ● Vitria BusinessWare™
Data/messaging management	● Frontec AMTrix™ ● NEON Integrator™	● Netscape ECXpert™ ● SAGA Sagavista™ ● SeeBeyond – eGate™ ● TIBCO ActiveEnterprise™ ● webMethods B2B™ ● Viewlocity™

Integration scope

Figure 6.8 Integration software market.

EAI offers the potential not only to share data within and across organizations, but to manage and use that data in an integrated way. These characteristics will be of enormous value to eMarketplace operators whose service offering includes collaborative supply chain planning. These applications are key enablers of eSynchronization.

Visions of a future eSynchronized world of interconnected marketplaces

Having explored the two concepts of eMarketplaces and a shift towards collaboration and synchronization, it is interesting to reflect what happens when they converge. Industry eMarketplaces have tended to ignore their interactions with other industries' marketplaces and the strategies of mini-hubs and horizontals. In reality, any vision of an eSynchronized world of eMarketplaces has to also consider the interaction of eMarketplaces. Key questions are: Who will be strong? Who will be weak? and Will the value of many of these changes leak out to customers in the medium to long term?

The discussion of these issues is still at a very early stage. If we are less than 2 per cent into the eProcurement market space then it is very difficult to make predictions about eMarketplace competition. However, it does seem fair to say that we can expect vigorous competition to secure the greatest value.

Inter-marketplace competition

In the scramble to set up industry eMarketplaces, many participants forgot to think through the issues associated with inter-eMarketplace competition. It is inevitable that this will be an issue as multiple large players from one industry encounter the consolidation of the demand or supply of another industry. The situation will be further complicated by the roles that investment bankers, software vendors, systems integrators and diversified multinationals play in multiple eMarketplaces. Many vendors will find themselves with very significant conflicts of interest as inter-marketplace competition accelerates.

It is not clear who will do best from an industry perspective. What is clear is that those companies who create the most liquid and profitable eMarketplaces and deliver on the 'onboarding' challenge will do much better than those companies that trust to fate. There is a general perception that value migrates to the customer over time. This is an assumption based on a perception that greater transparency of transactions and value always benefits the customer.

Scenarios for eMarketplace competition

With so much uncertainty still remaining around what the appropriate business models for the eEconomy will be, it is surprising that many organizations appear to be making the assumption that all required parties will come together, share their knowledge and build close relationships with each other without contention. Not only that, there seems to be unwavering confidence that all the required technology will be set up around the marketplaces, despite the known fact that vendors are quite often selling 'hope'.

We see a range of possible scenarios emerging, depending on the relative power of eMarketplaces and the ability of companies to consolidate. The framework in Figure 6.9 provides an interesting perspective on the possible scenarios for inter-eMarketplace competition. Most of the eMarketplaces seem to be assuming a scenario of 'everything works' – both inside their industry and at the immediate interface between their industry and others. This scenario is possible, but unlikely to happen across multiple industries. For example, the food and retail industries need to resolve local interactions between themselves around product coding before they can realistically take on broader cross-industry interfaces.

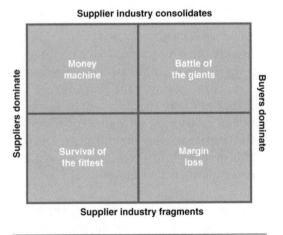

Figure 6.9 Scenario framework.

We have identified four sample scenarios to illustrate our concerns about inter-eMarketplace competition. These are:

- *Money machine*, where everything works to plan
- *Battle of the giants*: big inter-eMarketplace competition
- *Margin loss*: buyer-side wins
- *Survival of the fittest*: eMarketplace fragmentation back into company-to-company competition

Money machine
This scenario reflects the situation described above: that all parties will come together and cooperate and that the marketplaces will operate as planned by the industry players and their partners. The consolidated eMarketplace industry will hold the dominant position. They will set the standards and the prices, and be in full control. This sounds good, but will be very difficult to achieve.

Battle of the giants
In this scenario, both the supplier and buyer industries come up with consolidated industry solutions. This scenario tests how far vertical industry marketplaces will be prepared and positioned to compete against other well-organized eMarketplaces. For example, how will the chemicals or plastics marketplace respond to pricing pressures from the automotive industry? The potential 'unbalance of power' could necessitate very different approaches and strategies from those which are appropriate in the 'money machine' scenario.

Margin loss
In this scenario, the supplier industry finds that it has not come together in a clean, consolidated way, as in the two scenarios above. Add to this the pressure of a consolidated or powerful buyer industry and the resulting dynamic is very different from what many currently expect. Under these conditions, many suppliers will be vying for the business of dominant buyers who have the power to commoditize products and squeeze prices down.

Survival of the fittest
Finally, a scenario in which consolidated industry plays do not come to fruition for either the supplier or the buyer industry. The resulting dynamic can be likened to the traditional shopping centre situation where buyers have to come to the suppliers, the suppliers are competing directly with each other, and the survivors will be those that can last

through the lean times, find ways to differentiate themselves from the crowd and establish a healthy, dominant share in a niche market – the 'fittest'.

The implications for an organization's strategic positioning, capability requirements and required key change initiatives, are clearly very different for each of the above scenarios. Organizations must consider not only their own direction and the dynamics within their industry, but also that of the industries up- and downstream of them. They must ask what could possibly change the path for any, some or all entities. For example, regulator intervention or unconvincing value propositions could lead to the non-cooperation of certain parties. They must consider the drivers of these challenges or obstacles and make judgements on the likelihood of their occurrence. Organizations need contingency plans in such uncertain times – something many are neglecting to recognize and accommodate.

The 'Money machine' scenario, for example, assumes that participants in the industry do not mind sharing information about and the capabilities of their organizations because everyone is going to make money. In contrast, if organizations find themselves operating in a 'Survival of the fittest' scenario, the last thing they will be prepared to do is share information with their competitors.

Key challenges and lessons learned from eMarketplaces

eMarketplaces are still very young and underdeveloped. Our expectation is that most will fail, because they are not designed for success. eMarketplaces have been promoted as new mechanisms that will handle unbelievable levels of complexity seamlessly. This may happen over time – but not yet. eMarketplace failure is not a sign of a bad idea, just of bad or naive expectations. We believe that they have a big role to play in a future eSynchronized world. However, companies need to recognize that collaboration is more about the best partners and execution capabilities than the best and most original ideas.

Most eMarketplaces will fail to manage complexity

One of the major challenges for eMarketplaces is the effective management of high levels of complexity. Many have failed and will fail in their attempts to bring multiple parties together around the business model, financial structure, technical standards and business practices. The increasing number and changing nature of relationships adds another dimension to the complexities of eMarketplaces. When you consider that most eMarketplaces struggle with relatively simple commodity transactions, it is unlikely that there will be many that will be able to effectively manage the more complex categories of components, services and capital goods – wherein the vast majority of business spending lies.

As the restructuring of the financial services industry over the last two decades has demonstrated, the management of complexity requires the adoption of standards – global standards. Standards will serve to simplify all tasks, speed up the communications between all parties and enable the required levels of customization of goods and services in the open market.

Technology always costs more, benefits are lower and relationships are hard to sustain

Many eMarketplaces, in the rush to set up, have invested in relatively cheap off-the-shelf technologies, which have subsequently proven to be inadequate in terms of both functionality and scalability. The cost to fix problems and upgrade or change over to more appropriate technologies is often insurmountable, particularly as second round funding becomes more elusive.

Even the most successful eMarketplaces of the day are not realizing their projected benefits. Commission-based systems are sadly lacking as transaction volumes fail to mount. The key lesson here is that while transfer of product may be the ultimate conclusion of an eMarketplace interaction, the true value creator for the suppliers and customers, and consequently the eMarketplace, is more likely to be the information about the product and market.

The value propositions offered by the majority of today's eMarketplaces are in direct conflict with the latest relationship management strategies. Organizations have been focusing on developing closer, longer term partnerships with suppliers with the aims of streamlining processes, reducing inventory holdings and increasing agility and flexibility. The common eMarketplace model, which allows buyers to get the lowest possible price from the bidding suppliers, is unsustainable not only because of the absence of incentive for suppliers, but also because of the lack of consideration of factors such as quality, customization, and delivery timings – factors which could contribute more to a supplier's overall value than price.

Ideas are cheap, money is there, but 'Execution is the new king'

It is not difficult to develop the new ideas around eMarketplaces. There are large numbers of analysts' reports, Web sites, books and live examples from which you can draw ideas. If anything, the subject is characterized by the sheer volume of unimplemented good ideas. These are mixed up with a whole range of partially implemented 'bad' ideas.

We believe that the next phase of development of eMarketplace solutions will focus heavily on who can implement. Just think of a few of the challenges facing companies:

- The vast majority of B2B exchanges and marketplaces do not have any capability to 'close the loop' on fulfillment and realization.
- Proliferation of outsourcing, JV (joint venture) and Newco (new company) models requires integrated execution capabilities to capture the promise of speed and cost performance.
- Few entraprise and extraprise models today have synchronization capabilities outside their own firewalls.

Part of the challenge for traditional multinationals is finding people who understand the issues and enough people who can deliver the new solutions. In our experience, all traditional companies are dangerously short of both types of people.

Choosing the right partners

Choosing the right partners has been a constant theme throughout this book, and in the area of eMarketplaces it has special importance. Choosing partners to take on individual market opportunities is one thing. Choosing long-term partners that can help to deliver on the promise of eMarketplaces is a more difficult decision. It means testing out partners and jointly developing solutions. In a world of eMarketplaces the best will get better working with other star players, and the average will get worse. The challenge is to choose the right players to train and learn with.

Conclusion

Successful eMarketplaces will be well positioned to become the principal supply chain synchronization facilitators. We are already seeing agendas that combine eProcurement with eDesign, eManufacturing, ePlanning and eFulfillment across a number of industries. With the requirements for increasing collaboration and synchronization reasonably well understood, major work efforts are under way to build the supply chain planning and enterprise applications integration (EAI) solutions which will truly enable eSynchronization.

While many eMarketplaces are expected to fail, we still believe that they have a big role to play in the eSynchronized world of the future. Perhaps the single most critical success factor is the realization that collaboration and synchronization are more about having the best partners and execution capabilities than the best and most original ideas. The time is set for the best companies and eMarketplaces to exploit collaboration and synchronization technologies.

7 New information technology architecture for supply chains

Introduction

Applications and technical architectures do not usually grip the attention of senior executives. However, the days when such matters could be left to the chief information officer and the information technology department are long gone. In this chapter we explore developments in supply chain technology and new business models, such as netsourcing, that will inevitably result from latest innovations in software and hardware platforms, and explain the impact on supply chain management. Senior executives ignore this subject at their peril: it goes to the very heart of the connectivity and compatibility on which the future operations of all companies will depend.

The five key questions we address in this chapter are:

- Who are the winners and losers in applications software?
- What will be the hardware platforms of the future?
- What will be the impact of netsourcing applications?
- What will be the new models for outsourcing functionality to external suppliers?
- What are the implications for supply chain management?

Success and failure in applications software

The release of new packaged software has accelerated over the last few years, particularly in new areas such as customer relationship management, supply chain management and eCommerce interfaces. This proliferation has coincided with rapid standardization around a small number of applications packages and basic platforms such as Unix and Microsoft Windows.

Intense competition

Large numbers of software companies are scrambling to create a critical mass for their new products, and most software companies market their products beyond the product's actual capability. Failure to secure sales at an early stage of product development spells early death. This is often referred to as selling the 'gap', or selling capabilities from PowerPoint presentations and building the product later.

Each application area is seeing the emergence of just three or four winners – those that stand out are Ariba, Oracle and Commerce One for eProcurement and Siebel, Vantive and Oracle for customer relationship management solutions. In the software applications market, the market leader typically achieves a 50 per cent market share, 10–12 per cent goes to the second in the market, and the third earns about 8 per cent. It is not clear whether this pattern will hold true for these new applications – but it is likely. The losers in these software wars either go bust or are swallowed up by their rivals, and the lucky few, such as Aspect Technologies and Trading Dynamics, have earned a great deal of money in the swallowing process.

As the three or four winners emerge, they work hard to make their software compatible with major technical infrastructure platforms, including Unix, Windows NT and Windows 2000. They also develop interfaces to connect to the major enterprise resource planning products – SAP, Oracle and JD Edwards – and with other established application packages, such as those from Siebel and i2 Technologies.

Finally, the software application providers fight hard through development, acquisition, alliances and marketing to fill gaps in their products and capture 'white space' or open areas of functionality not currently satisfied in the marketplace. Examples of white space include industry-specific software and new areas such as wireless application protocol (WAP) for mobile phones.

Competition in the software marketplace has become significantly more intense over the past 18 months, an inevitable consequence of the growing gap between the rewards for success and the penalties for failure. For example, there is already general acceptance of who the three to four winners in eProcurement are, despite the fact that penetration of this marketplace has only reached 1 per cent, and few companies have completed more than 10 per cent of their programmes.

Early on, those leading eProcurement solution vendors maximized their 'mind share' among executives and opinion influencers – market analysts such as Forrester Research and GartnerGroup and major consultancies such as McKinsey and Accenture. The early days of eProcurement sales were remarkable. Never had so many presentations been so well received by so many companies with so little actually being bought! Nevertheless, a small group of companies managed to gain access to huge amounts of capital from the financial markets and were able to develop their products and acquire their competitors' employees and technology. With few developed products and global staffs of 300 employees, some of these software companies were able to command market capitalizations up to four times greater than those of major software companies, such as SAP.

Emerging winners and losers

The likely winners and losers in the software application market are now becoming clearer, although inevitably there will be some surprises. The applications maturity map (see Figure 7.1) shows the leading package vendors in each functional space and the funnel effect, whereby products become more accepted and more mature as five or six leaders are reduced to two or three.

Companies such as Sun Microsystems and Cisco Systems, together with the major systems integration consultancies, will directly affect whether individual packages succeed or fail by deciding whether to endorse them, influencing their development strategies and training their own employees to implement them. Scalability is the key to success. The more people – both internally and more especially externally – that support a vendor's package, the more likely that package is to succeed.

Until recently, many dotCom businesses established their own eCommerce strategies, designed their own technical architectures, and managed their own integration and deployment of solutions. Today, however, the winners in the dotCom world are partnering up front, leveraging the expertise of technology leaders and utilizing best-of-breed solutions. Accordingly, Sun Microsystems has launched a best practice programme that leverages its accumulating knowledge and experience in helping customers to 'dotCom' their businesses. Called iForce, the initiative provides a road map that outlines four major steps that customers must take when dotComming their businesses – defining the business strategy, building the architecture, developing the integration

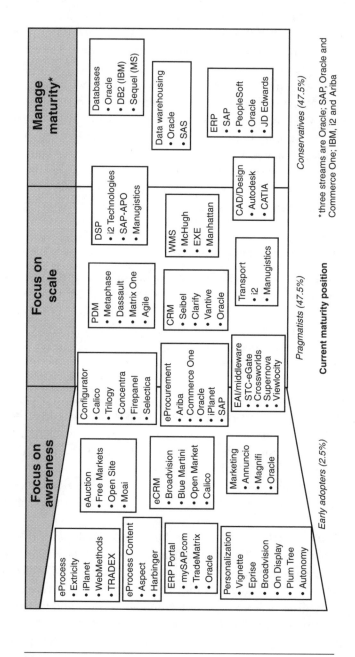

Figure 7.1 Applications maturity map.

strategy and deploying and managing the solution. iForce provides organizations, from mature enterprises to start-ups, with the products, services, solutions, road maps and partnerships they need to plan, implement and maintain a dotCom strategy successfully. It provides customers a structure for organizing the complex process of creating a dotCom, while at the same time exploiting Sun's expertise each step on the way.

To augment its implementation capabilities, Sun has established iForce Ready Centers[SM] in California, Paris and Tokyo, where its dotCom experts join with independent software vendors and implementation partners under one roof to help customers create proofs-of-concept for their businesses. By assisting the customer with everything from brainstorming the technological options for creating a wireless infrastructure to actual pilot programmes, and by providing guidelines for the design, development and deployment of dotCom services on the Sun platform, the centres reduce risk and time-to-market. Their solution sets aggregate established, scalable and simple to customize best-of-breed applications, follow open standards and bring a proven blend of software, hardware, networking, financing and consulting services into a single, manageable relationship.

Making the right applications decisions

In a world of scalability, connectivity and standardization, senior executives must realize that they cannot afford to buy into non-standard packages.

Major business partners will concentrate their integration and connectivity efforts around the market-leading packages, whose vendors will have both the capital and the incentive to ensure faster development of their functionality. At the same time, the major systems integrators will have a stronger incentive to build interfaces to, and adaptors between, the leading packages and will be reluctant to invest in interfacing with the also-rans. Similarly, information technology professionals will be unwilling to stay with a company where they have to learn and develop marginal skill sets and it will be difficult to find suitably trained external resources to support a unique applications architecture.

The emergence of the three major streams of packages (Oracle; IBM, i2 Technologies and Ariba; and SAP, Oracle and Commerce One) has major implications for vendors and multinational companies. Vendors must either back a single stream or make their products compatible with all streams. Companies, on the other hand, can now increasingly rely on external parties to build the critical interfaces between packages, as long as they stay with the major players. The days when an information technology department spent up to 50 per cent of its time building interfaces between unique combinations of custom and packaged software are over.

The importance of data and ERP

Enterprise resource planning packages and databases will remain the bedrock of companies' information technology infrastructures. Most companies have yet to achieve the goal of a single ERP system. Partly because of acquisitions, partly because of geographic rollouts and partly because of unit autonomy, it is not uncommon to find multiple packages and multiple instances of ERP systems within a single organization. It is unlikely that those organizations that wish to be effective in an interconnected business world will be able to do so with fragmented systems. Rather, there must be a single ERP system across the entire company – perhaps an 'ERP Lite' version limited to a simplified subset of functionality. Increasingly, complexity will be managed through eCommerce and bolt-on applications packages.

Data will also become a more challenging issue than it is today. Companies can hide their fragmented and complex systems when they do not have to link to external partners. However, growing connectivity and collaboration between organizations will require a level of discipline and quality of data that most organizations have yet to even think about. For example, if a company creates poor demand or supply planning forecasts for three months in a row, it will have a difficult time maintaining alliances with best partners. Quick fixes for data issues, such as using spreadsheet wizards that aim to resolve confusion, will be unacceptable in a world of exchanges and eMarketplaces.

Most organizations have a significant change management task first to determine how to standardize around a single ERP system and data warehouse and then to support their people to maintain the standard processes and clean data. In the final analysis, this is not an issue of technology but one of executive leadership and people.

Hardware platforms for the future

The consolidation of hardware and technical infrastructure platform companies has been even more dramatic than the consolidation of software applications. In this arena, the battle is between three major platforms represented by the Unix/Java-based solution led by Oracle and Sun, Microsoft's NT and Windows 2000 solutions, and a tailored IBM solution. Competition is based on trade-offs between cost, scalability, ease-of-use and levels of reliability.

CASE FILE

Within days of its October 1998 launch, Egg, Prudential's online financial organization, received 1.75 million hits on its Web site inquiring about its three initial products: a savings account, mortgage and personal loan product. This unexpected level of success has continued. Egg now processes between 800 and 1 000 applications each day and has a sustained ability to handle some 400 concurrent customer transactions.

Egg's approach differs from that of the traditional banks that offer inflexible off-the-shelf products. Instead, it tailors products to meet individual customer needs. This requires a system that can capture the information to enable analysis of any one customer at any given time so that Egg can manufacture a specific product based on an individual profile.

Egg's initial architecture failed to match its aggressive aspirations for rapid growth and the lightning-fast take up of its products and services. It needed an architecture that could process unpredictably high volumes of customers and transactions during short time frames. Its original architecture could not handle a volume of customers and transactions that was 10 times greater than expected. As a result Egg switched to the Sun platform, which was scalable and robust enough to ensure that Egg could deliver its key business results.

The problems of scalability have been well illustrated by the experiences of many dotCom start-ups, where a combination of poor architecture, naivety and lack of basic internal management processes has led to spectacular crashes and losses of value. Indeed, some hardware vendors have adopted a sales strategy of approaching dotComs only when they have had their first significant crash. These vendors have found that they can then sell their products at list rather than discount prices once the dotCom's management and financial backers have experienced the real challenges of scalability. Smart companies are those who will back a scalable architecture with effective data centres at an early stage in their development.

Major software vendors have an incentive to ensure that their products work on each of the major platforms. However, trends within industries will influence decision-making about hardware platforms. Industries with high levels of transaction processing, security and reliability, such as financial services, will tend to favour Sun and IBM solutions. Conversely, many chemicals industry companies favour Microsoft platforms, which is likely to influence the design and technology of industry exchanges.

The selection of hardware and technical infrastructure platforms should be driven by cost, scalability and access to industry-level solutions. Executives should ensure that their selected solutions will not disadvantage them relative to their competitors. They should also take care not to waste time and money and risk long-term problems by backing multiple solutions in different geographies. As the hardware and software market evolves, increasing use of shared data centres to service multiple geographies, greater use of mobile computing and standardization of desktop solutions are likely to emerge. A rapid convergence around simple standards across the entire organization will simplify the information technology workload, reduce service costs and increase opportunities for procurement efficiencies.

Cross-enterprise standardization will have its opponents, often being people such as country managers who have little interest in or knowledge of information technology and who fear a loss of control over technology platforms, a reduction in supervisory and procurement responsibilities or discovery of poor decisions made in the past. There are two common approaches to handling these obstacles. A strong commitment by the executive committee to adhere to common standards will require that employees who want to break standards justify their case beyond any doubt. The other is to adjust human resources and financial controls to allow local or functional management control of standards – but their adherence should be measured.

New approaches to netsourcing models

Netsourcing is a business model and delivery mechanism that allows a third party to provide and manage a common business solution over an electronic network for multiple clients. Although netsourcing is a relatively new phenomenon, it is not unlike the traditional approach to bureau services.

With netsourcing, it is the provider, rather than the customer, that buys and owns the hardware and software required for a particular application or solution. The customer buys functionality, services and solutions 'by the drink'. If the customer's business grows, the netsourcer charges more to handle the greater volume of business; if it contracts, then charges decrease. Such a model offers a number of attractions. Information technology costs become more variable and high capital costs are avoided. Standardization of processes and technology, controlled by the netsourcer, is imposed on the buying organization. Staff numbers are reduced and the company gains access to people with scarce information technology skills, as a result of higher pay and greater equity rewards.

The disadvantages of netsourcing are mainly loss of control, loss of long-term capabilities and vulnerability to future pricing changes. However, most of these can be controlled by good contract management and clear policies on software upgrades and exit strategies.

Netsourcing will be used for an increasingly wide range of business solutions, from enterprise resource planning to order management, human resources and supply chain planning. There are two major models. The applications solution provider (ASP) provides hosting of an application such as an enterprise resource planning system, while the business solution provider (BSP), which may use an ASP to host its services, provides an entire business solution such as order management or human resource services (see Figure 7.2).

Netsourcing models are still in their infancy; however, a few are up and running. Asera, a very high profile Silicon Valley start-up, initially positioned itself as a netsourcer of electronic order entry processes. It has secured 20 major clients and an impressive list of strategic investors in a fraction of the time normally required because of the power of its best-of-breed netsourcing model.

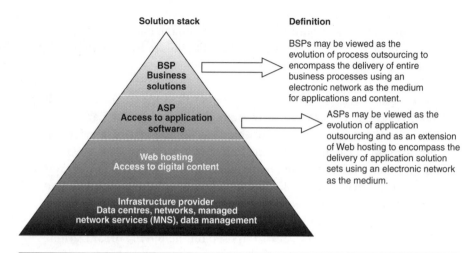

Figure 7.2 Netsourcing solution stack.

CASE FILE

Launched in September 1999, Asera provides Internet solutions for the business-to-business demand chain. It provides netsourced, eBusiness solutions that allow companies to incorporate new functions and features to keep pace with their specific business needs. Their business model and delivery mechanism, in which a third party provides and manages business solutions over the Internet, enables clients to incorporate new features incrementally and pay only for what they need and use. Clients can select from order management, content management and community services, such as message boards and industry forums, and maintain control of their eCommerce strategies through personalized Web sites, for which they pay an initial activation fee based on monthly usage instead of large, up-front software licensing fees.

The Asera eService is highly configurable and can incorporate innovations rapidly to improve its clients' ability to respond to the evolving needs of their channel partners and end customers and enhance sales volume and customer loyalty. The eService can be tailored to include a client's unique brand characteristics, content and specialized processes during an initial 60- to 90-day 'activation' period.

Initially, Asera is focused on providing channel solutions to the electronics, high-tech, communications, financial services and chemicals industries. However, as a flexible business-to-business utility for the demand chain, the new offering has relevance to virtually any industry that relies on brokers, resellers or distributors, as well as to those companies that wish to provide their key customers with personalized portals to access their organizations.

New models for outsourcing functionality to external suppliers

The demise of information technology and other functional departments, the emergence of a broad range of new software functionality and the dramatic increase in the scalability of hardware platforms have led to outsourcing business functions that many organizations have traditionally regarded as strategic and closed to outsiders. This shift demands a new approach to the ways in which outsourcing relationships are conceived and executed.

Traditional outsourcing has generated a great deal of bad blood between partnering organizations and has generally failed to fulfil expectations – many organizations are now sceptical about continuing to use third-party outsourcing models of the past. Despite this situation, the current state of back office activities in many multinational organizations is no longer sustainable. Many of these organizations have been through at least 20 years of significant cost cutting, business process re-engineering and efficiency programmes. As a result, they have recruited fewer resources, failed to retain fast-track managers and focused attention on the importance of business units rather than functions. Support functions such as information technology, logistics, human resources, procurement and finance have been downsized gradually, and many have lost their best people because of an inability to show a clear and attractive career path. Such organizations are nearly beyond the point of being able to reinvent their back office functions. While they may have the management talent to do so, they lack a strong business case to redeploy lost talent to internal functions or recruit new resources.

Therefore, a rapid and sustained move to outsource the remainder of such functions and processes to preferred suppliers is inevitable. However, these organizations will not be content with arm's length, third-party outsourcing. In many cases, they will want equity in the outsource ventures in order to maintain control (or at least retain influence), and to gain access to performance data.

Most traditional outsourcing relationships have failed to create a win–win situation because both sides have tried to tie down metrics and costs too early in the relationship. Indeed, in many cases, there has been no relationship until the contract is signed, and only a very difficult one thereafter. Experience has shown that understanding and trust need to be developed,

CASE FILE

BP Exploration Operating Company Limited (BPX), the Aberdeen-based arm of the oil giant, British Petroleum, operates more than 20 fields in the North Sea, and its core activities are finding, extracting and selling oil. In the early 1990s, against a background of maturing fields, increasing operational complexity and rising costs, BPX had to devise new, more cost-efficient ways of doing business. The aim was to reduce the company's spending substantially in the short term. Although the use of contractors was established practice in the oil community, the concept of contracting out business-critical accounting services as sizeable and complex as BPX's was untested. However, it offered the opportunity to transfer resources to a specialist third party contracted to deliver a flexible, responsive and reliable service at a predictable cost in a long-term partnering arrangement.

Under the agreement between BPX and Accenture, some 320 BPX staff from six locations transferred to a new entity, Accounting Services Aberdeen (ASA), in July 1991. Tight definition of roles, responsibilities and service levels, coupled with continuous performance measurement and a commitment to train and develop staff, underpinned the agreement. While BPX retains control of policy setting and interpretation of information, ASA handles all accounting functions. Key aspects include: joint venture accounting; provision of management information; tax reporting; and the preparation of group and statutory accounts; as well as processing and paying between 12 000 and 13 000 invoices a month. The contract is managed by a Joint Review Board of three representatives from each company who have shared business objectives and a common understanding of the critical success factors.

As a result of the success of the first phase of the arrangement, the contract was renegotiated one year early and renewed from June 1994 for a period of five years until June 1999. Unlike Phase I, which was a fixed price agreement with a guaranteed reduction, Phase II was an incentivized cost-plus arrangement based on a target fee reduction of 5 per cent per annum, but subject to the same change parameters. Over- and under-spends against target are shared 50/50. The scope of the arrangement was also extended in 1994, and in 1996 it was announced that accounting for all BP companies in North America would be contracted to Accenture.

After all joint venture costs, the reduction in BPX's accounting costs reached 40 per cent within the first six years of the formation of ASA. Measurable savings in audit fees and cost of funding resulted in 7 per cent further savings. Other benefits achieved include:

● Simplified and standardized accounting activities
● Transformation of BPX into a more service-driven organization
● Provided flexibility to respond rapidly to changing industry requirements

Customer satisfaction: BPX managers surveyed rated the outsourcing service as very good (options: poor, requires improvement, good, very good, excellent). Partners and suppliers rate service good and very good respectively.

often over as long as six months, before agreeing to service level agreements and key performance metrics, if these relationships are to be sustained.

New types of fourth-party outsourcing arrangements are emerging for back-office functions. With third-party arrangements there is a high probability of a win–lose or lose–lose scenario because the arrangements are arm's length and contractual. With Fourth-Party Logistics™, a concept created by Accenture, there are cross-holdings and, potentially, shared employees, so that a failure by one party affects both parties.

Fourth-party offerings have been topics of discussion for about three years, and they are just beginning to gain acceptance. Models such as Optimum Logistics and ShipChem for logistics and ePeopleserve and Exult for human resources are a few of the latest examples, with more expected on the horizon. The latest 4PL arrangement entered into by Accenture involves a global communications industry service provider.

CASE FILE

Unisys, an electronic business solutions company whose 36 000 employees help customers in 100 countries apply information technology to overcome challenges of the Internet economy. Exult is delivering Unisys an eEnabled global human resources solution and redesigning and managing a number of Unisys's human resources processes and services while Unisys retains accountability for its human resources strategy and planning and for the development and management of strategic human resources programmes. The comprehensive solution builds on Unisys's substantial investment in human resources process excellence and Web enablement and will provide Unisys employees with a self-service environment and access to a wide range of processes through Exult's integrated approach. As a result, Unisys anticipates a reduction of some $200 million in its current human resources expenditures over the life of the seven-year contract.

Implications for supply chain management

Few supply chain executives will want to enter into detailed discussions on supply chain collaboration and synchronization with strategic business partners without being certain that their own resources have the tools and processes needed to deliver an acceptable demand or supply forecast. Executives who believe that the supply chain should be a key competitive weapon rather than a millstone around their corporate necks need to appreciate several considerations:

- Executive ownership and understanding of technology architecture decisions and implementation issues increasingly determines which companies are the best value chain partners.
- Most multinationals will need a single, common enterprise resource planning system that is integrated with best-of breed-applications and Internet technologies.
- Data is the lifeblood – those who manage it should make sure they understand the need for it to be clean and accurate.
- Companies selecting application packages must follow the crowd or be brave enough to direct it. Unique solutions will increasingly be unsustainable and painful for those who attempt to impose them on their organizations. Smart companies recognize the economy and sustainability of being part of the mainstream.
- Unjustified process fragmentation will become less tolerated. In the future, doing business will involve processes that are common both within and across organizations.
- Supply chain integration, collaboration and synchronization will require a new approach to sharing data. Companies will have to share good data with minimal human intervention and rapidly respond to changing business situations.
- Security and robustness will become increasingly important issues. No one will want to partner with a company that cannot protect its systems or sustain high levels of systems availability.
- Outsourcing and netsourcing will become increasingly common. Even those companies that do not outsource capabilities will have to work with the outsourced and netsourced partners of their customers and suppliers and will have to keep up with their innovations and new approaches to doing business.
- Successful supply chain management will require fewer and smarter people as data on cross-company and cross-industry flows become more available. Companies are more likely to need small numbers of PhD level mathematicians than large numbers of low-level planners.
- Preparing the organization for the impact of new technology architectures on the supply chain is a tremendous change management task. Now is the time to begin preparing.

Conclusion

Technology has the potential to create a new level playing field within companies, across and between industries. In this new – and initially seemingly unnatural – phase of interconnectivity between companies, technology can be a key competitive weapon, but only if it is used intelligently. The leading companies of tomorrow have already designed an approach to information technology architecture, data and enterprise resource planning that will make them formidable value chain partners and competitors.

8 New ways to deliver eWorking and continuous innovation

Introduction

Internet technologies are now creating the opportunity to increase white-collar productivity through new ways of eWorking. The goal of eWorking is to eliminate non-value-added activities, support employees with relevant information and put in place processes and metrics that improve employee performance. eWorking leverages Internet, intranets and other electronic technologies to develop employee portals, standardize processes, create new approaches to knowledge management and enable continuous innovation, while streamlining processes and eliminating waste from the system. Although these ideas are common in newer companies such as Cisco Systems, Oracle, Sun Microsystems and Microsoft, they are only beginning to appear in traditional organizations like Dow Chemical, BP and GE.

To understand the opportunity for eWorking, think about the proportion of your time, and your colleagues' time that is spent on what you consider to be non-value-added activities. Write the amount of time down and compare your estimate with those of your colleagues. Over the last few years, we have asked this question of hundreds of executives and employees. On average, they estimate that 30 per cent of their time is non-value-added, although responses have ranged from an unrealistic zero per cent to a staggering 100 per cent from one disgruntled employee. Assume that our average roughly applies to your organization, and imagine that you could cut it by half. That 15 per cent would translate to an extra hour of productive time each day. eWorking not only helps organizations to gain that one hour per day, it also makes it twice as productive. Effective management of the increasing clockspeed, complexity, number and criticality of business relationships requires employees to make significantly improved decisions. The less time spent searching for information the better, particularly in a supply chain environment in which real-time, informed decisions are required at all levels of the organization. The potential impact of eWorking on supply chain management is enormous.

In this chapter we explore the ways in which companies can provide systems and processes to help their employees become more productive and fulfilled as individuals and as teams. We address six key questions:

- How are companies using employee-centric portals to enable eWorking?
- How are companies using the latest Internet and communications technologies to improve employee and investor communications?
- How are companies standardizing policies, procedures and processes?
- How are they linking employee portals to new approaches to knowledge management?
- How are companies building a culture of continuous innovation?
- How are companies achieving greater value from their project planning and resource allocation processes?

Employee-centric portals

Savvy organizations are beginning to utilize the technology behind companies such as Amazon.com and Yahoo! to develop company-level portals that run inside their firewalls. These portals, accessible through employees' desktops, provide relevant information that extends beyond typical industry news-feeds and company financial reports. They are evolving into one-stop shops for all the information needs of an employee. With quick and easy retrieval functionality, these employee-centric portals are becoming the channel through which information is disseminated throughout an organization.

From simple to-do lists to transaction data in enterprise resource planning (ERP) systems, the value derived from these portals is manifold. Designed to source data from a small number of databases, portals simplify the standardization of information across the entire organization. They encourage the dissemination of innovative ideas as new ideas and tools developed in one part of the organization can be shared with and rolled-out to the rest of the organization rapidly. A new philosophy of information management in which relevant information is made available to employees with reduced search time is spurred by portals, and with a single point of reference, new employees can be assimilated more easily. Finally, to support change initiatives, portals enable technology advancements and new business processes to be shared, benefits communicated, and training simultaneously rolled out throughout the entire organization.

In the race to develop sophisticated solutions, many companies are establishing programmes to ensure that every employee has access to a computer and is comfortable using it. Ford, for example, recently set up a programme whereby every company employee could sign up to receive a personal computer, printer and Internet access for only $5 a month. The aim was to give every single person in the company, and their families, the opportunity to develop their own skills and learn the latest technology trends. Delta Airlines has undertaken a similar scheme to move its entire 70 000 employee workforce online.

CASE FILE

Cisco Systems is a well-known leader in implementing eCommerce and Internet capabilities. It has a comprehensive eWorking programme that includes Customer Care, Internet Commerce and Supply Chain Management. These programmes achieved financial benefits of $770 million during 1999.

Cisco saves over $75 million, or $2500 per employee annually, as a direct result of its implementation of workforce optimization applications. These applications, based on Internet technologies, enable employees to focus on the core value of their jobs. Measured in terms of revenue per employee, Cisco employees are almost twice as productive as its closest competitor in the networking industry. The organization's interactive intranet Web site, Cisco Employee Connection, allows employees to access information on demand and perform business transactions and administrative tasks without administrative support.

The company's range of intranet tools and its streamlining and standardization of processes have helped to provide scalability through automation and self-service. Components include a personalized portal, an employee administration directory, recruiting and hiring communications, payroll and expenses reporting, compensation review, training system registration and completion, employee communications and customer and account information. eWorking benefits realized by Cisco include administrative cost reductions, increased employee productivity and employee satisfaction, leading to a reduction in employee turnover (see Table 8.1).

Table 8.1 Cisco's 1999 financial benefits achieved (source: `http://www.cisco.com/` – multiple pages and white papers).

Benefits	Measured value
Improved employee communications	By shifting to a Web-based communications model, Cisco has recognized over $16 million in employee productivity.
Improved expense reporting	Savings of $6.1 million were projected to be achieved by the year 2000 with a reduction in payment time by 80 per cent to four days.
Decreased recruiting costs	Increased effectiveness of the Web résumé system saves Cisco $3 million annually.
Decreased employee turnover	Turnover dropped from 11.5 to 8.6 per cent partly due to improved recruitment approaches, employee communications and employee satisfaction.
Optimizing existing head count	With automation, the HR department avoided increasing headcount by 27 employees, saving Cisco over $2.7 million annually.
Training	Immediate online training enables support and sales staff to train quickly, saving Cisco over $24 million annually.
Online policy manuals and company directory	Online policy manuals and directories save Cisco more than $3 million annually in distribution and productivity for providing immediate access to accurate information.
Added value of information and learning	A figure of 1 per cent of revenue has been placed on the value added through increased productivity as a result of improved information and learning: $110 million per year.

CASE FILE

Accenture developed a Global Markets Portal to provide a central Web-based repository from which Accenture executives access information to support sales efforts. The portal provides detailed information about clients, including personal facts and a history of their relationship with Accenture, and automatically generates 'client fact packs' in minutes. It contains the latest external and internal thinking on hot topics, thought leadership and innovative engagements. The information is extracted from hundreds of carefully selected Web sites, news-feeds, and executive-focused Accenture Knowledge Exchange items. Executives can customize their own home-page – a 'My Page' – to display information about specific clients and industries of interest, stock prices, world time and even the weather. The portal also provides access to tax and legal advice with a click of the mouse.

The Global Markets Portal provides partners and client teams with a common go-to entry point. Using advanced search engines, discussion boards, project workrooms, comprehensive client and opportunity databases, and quick paths to the Internet, the portal offers an intuitive, rapid way to access information, share ideas, get answers and communicate innovations.

The core net-centric architecture uses Microsoft's DNA technologies. Plumtree, one of the leaders in the corporate portal market, provides browse, search and personalization capabilities. GetAccess from Encommerce provides the portal's single sign-on capabilities. Legacy integration with Lotus Notes databases is based on Lotus Notes Domino technology.

CASE FILE

BP Gas & Power teamed with Accenture to release eDesk, its Web-based portal that gives employees access to daily information and communication tools. eDesk users can customize home-pages to display the features and functionality that support their individual productivity, including email, news-feeds and competitor information from external agencies, weather reports, internal information required for job performance and a virtual file directory to access data from all over the BP business.

Applying eDesk technologies to manufacturing operations, BP Chemicals Chocolate Bayou Works (CBW), in conjunction with Accenture, has developed an employee intranet portal – My.CBW Portal – to improve employee productivity and lower the cost of managing information.

CASE FILE

MSWeb, Microsoft's Corporate Portal, aims to provide access to general-purpose information and knowledge about Microsoft, the industry Microsoft competes in and the employee services, applications and content sources relevant to the individual employee. It has been designed to be a fast, relevant and reliable channel for internal communications and strategic information and a comprehensive guide to key internal and external events. MSWeb is to be a single point of access to knowledge sources across the corporation. Employees can customize MSWeb, giving them the ability to modify the content and visual look of the MSWeb pages. They can also subscribe to targeted news and features to be received in mail or displayed on the MSWeb homepage. MSWeb also provides access to a number of other productivity tools including 'To Do Reminder Lists', search and retrieval functionality and tickers. The MSWeb already handles volumes of around 10 000 unique IP hits per day, 30 000 per month, 1.5 million page requests per month, and 2 000 search queries per day.

Broadcast, conferencing and collaboration technology

A new range of Internet- and intranet-enabled technologies are emerging which are helping to realize many of the promises of the Internet. They are enabling and facilitating processes such as real-time, broad-reach, simultaneous communications; collaboration and teamworking between geographically dispersed individuals or groups; and real-time project management. Leading companies are using these technologies to create new information speedways linking employees across dispersed business units and organizations. Broadcast, conferencing and collaboration technologies, such as WebCast, eRoom and NetMeeting, are being distributed to, and accessed via every desktop allowing instant access to groundbreaking news and virtual teamwork opportunities.

WebCast technology enables the transmission of an event, either live or recorded, over the Internet. The event may be broadcast publicly or to a targeted audience at their personal computers, enabling companies to reach employees, investors, analysts and other experts.

eRoom, by eRoom Technology, Inc., is a virtual Web-based workplace for business projects with provisions for team documents, discussions, voting and managing milestones, issues, tasks and communications. Members, both internal and external, benefit from access to the project material from anywhere at anytime. The tool automatically notifies team members of project updates and organizes project work in one place to capture intellectual capital as it transpires. When the project is complete, the intellectual capital can be viewed, searched and used as best practice material for the future. More than 250 companies including AT Kearney, Bausch & Lomb, Compaq, Cisco, Deloitte Consulting, EDS, Ernst & Young, Diamond Technology Partners, Hewlett-Packard, KPMG, Pfizer, Towers Perrin, Siemens and 3Com now use eRoom.

Microsoft's NetMeeting collaboration software provides a comprehensive suite of data-sharing and conferencing tools. It enables users to collaborate with any number of associates in real time. Information from one or more applications can be shared, graphics exchanged or diagrams drawn with an electronic whiteboard, messages can be sent, meeting notes recorded and items actioned with a text-based chat programme. Files can be sent to other meeting participants in the background. With a microphone, sound card and speakers, NetMeeting users can talk to each other over the Internet or corporate intranet in real time. With a video capture card and video camera, users can send and receive images for face-to-face communication or to run product demonstrations. Finally, the software allows one user to take control of another user's desktop remotely – an interesting capability that could be used, for example, as a technical support tool.

The impact of such advanced communication technologies will be felt at all levels of the organization. C-level (chief executive) skills will resemble those found more commonly in TV studios than in business schools. Executives will be expected to possess a telegenic presence and the ability to respond instantly and effectively to questions and queries

CASE FILE

Dow has empowered its employees with real-time information sharing and collaboration by incorporating many electronic communication advancements into its infrastructure. For example, more than 200 employee NetMeetings take place on Dow's system each day. Dow employees from all over the world can share information and tackle business issues in real time. This increased communication power enables Dow's global businesses to find the information they need to support analysis and decision-making quickly and easily. The company's battery of global electronic communications technologies includes:

- Standardized workstations for all employees to give the entire organization access to the same resources.
- NetMeetings to enable geographically dispersed employees to conduct virtual, productive work sessions.
- Videoconferencing to provide face-to-face dialogue for Dow and its global customers and partners, and technical support for offices in remote locations.
- Video streaming to employee desktops to announce company news to all employees simultaneously and conduct company-wide learning sessions.
- Satellite communications to deliver large-scale presentations to key stakeholders with excellent resolution.
- Webcasting to communicate with thousands of geographically dispersed people through an interactive online environment.
- Web video to view archived video presentations online.

Dow has also embarked on a programme to provide employees with the connectivity and convenience of a home office irrespective of their location. The DowNET 2001 team aims to integrate audio, video, telephony and computer capabilities into one tool to maximize employee productivity and satisfaction. The project also aims to provide future capabilities such as unified messaging, personal computer telephony, wireless communications and a 'follow-me' phone number.

CASE FILE

Cisco uses Webcasts, known as 'CiscoCasts', to deliver information directly to its employees' desktops using audio, video and interactive capabilities. Operating over a wide area network, Cisco uses this technology for its quarterly company meetings, allowing remote employees to view presentation slides and real-time video to simulate the experience of being a member of the live audience. Cisco also uses CiscoCast to conduct small meetings with employees from different locations and archives many of its video broadcasts for employees who are unable to log in for the live sessions.

and may frequently find themselves in company studio sessions to conduct global broadcasts, facilitate discussions, conduct question and answer sessions and respond to employee concerns.

These and other communications technologies carry the added benefit of helping to reduce negative activity, rumours, political issues and misinformation by providing the opportunity to communicate good information to all employees simultaneously. At Microsoft, following the verdict related to antitrust charges, Bill Gates and Steve Balmer were immediately able to share their reactions to the ruling with every Microsoft employee. Joe Forehand, the CEO of Accenture used a Webcast to communicate the results of an arbitration ruling regarding Arthur Andersen to 70 000 Accenture employees.

The benefits being gained by the leading organizations that are implementing eWorking practices are impressive. However, the advancements of leaders, while often inspiring, cannot simply be transplanted into other organizations. Attempts to do so will more likely result in employee resistance. Portals and portal functionality must be tailored to suit the particular company. But as more and more companies recognize the need for a unified approach to information dissemination and collaboration opportunities for virtual teams, eWorking is fast becoming a way to foster an environment of flexibility and innovation.

Standardization of policies, procedures and processes

The power of an employee-centric portal lies in its ability to offer a single standardized channel through which an entire organization can be serviced – with little tolerance for exceptions. Standardization, therefore, is the basic building block that enables the development of eWorking capabilities. The initial barrier to success in this unique channel is that most organizations' employees work in fragmented ways. This is especially true for multinational corporations, which suffer from geographical dispersion. Employees, used to different policies and procedures, may become unsure about which are valid and which have been replaced. Without standardization of policies, procedures and processes and employee training, it is virtually impossible to move an organization to one portal and subsequently unleash the value of eWorking.

With globalization, it is important for efforts to be coordinated worldwide and standardization to be oriented around the lowest common denominator. Simplifying and minimizing the number of instances of any given system or process – and using a company-wide software package or vendor – supports this effort. This results in a platform through which all company information can flow, eliminating the expensive and wasteful allocation of resources required to integrate disparate systems. An eWorking effort of this sort inevitably requires discipline to conform to template structures that minimize customization and exceptions.

Two key imperatives exist for successful standardization within an organization. Standardization requires support from the executive level, not only initially to establish the standardized processes, but also to ensure proper and ongoing adherence to standardization principles. Without this high-level commitment, individual departments will focus on their own needs, resulting in an uncoordinated effort. Secondly, standardization requires information that is maintained at its source. Information is best maintained by the people who are closest to it. For example, personnel contact information should be maintained by the employees themselves. When designing standardized processes, an organization should ensure that the information is captured in its most raw form by the people who are closest to it to guarantee the validity and applicability of the data.

CASE FILE

In May 1999, in an effort to reduce its overall operating costs by $1 billion, the Oracle Corporation embarked on a strategy that focused on the consolidation of global information technology. Oracle put its own customer relationship management products and philosophies to work – streamlining and standardizing business processes and procedures, and automating routine business functions using self-service Web applications. The company has moved from an administration process model to a self-service one that has empowered employees and simplified routine business functions. The source for gaining business intelligence has shifted from processing transactions to comprehensive global information. Already Oracle has reduced its IT costs by around $200 million, made significant improvements to operations and sales-related activities yielding savings of around $550 million, and streamlined and standardized its business practices around the globe. Oracle's eProcurement activities have already returned savings of around $150 million. The self-service model and automation of simple business processes, such as processing expense reports, purchase requisitions and travel bookings, is saving the company around $100 million per annum. Benefits associated with the increased productivity of employees as a result of standardization and eWorking are also being realized.

As an organization implements standardization principles, its fragmented entities begin to work in harmony with a common understanding of business processes and a confirmed belief in mutual goals. eWorking becomes especially important in an era of mergers and acquisitions, where organizations are often faced with the dilemma of having to pick and choose the best processes and systems within each company. These decisions, more often than not, simply result in the stronger organization being selected. By standardizing itself internally, a company can prepare itself for future growth with the ability to integrate merged operations and assimilate new employees easily so they can quickly focus on their key responsibilities.

Employee portals and knowledge management

Knowledge management has always been a mythical goal, something on a PowerPoint slide. Most agree on its importance, but few have been able master it. Most organizations operate their knowledge management departments like libraries, gathering and cataloguing seemingly endless amounts of information, with no way of synthesizing the knowledge captured. Running a search for a specific item of information often results in the identification of thousands of pieces of potentially relevant data, requiring the employee to wade through enormous amounts of information.

To be effective, knowledge management must be more than a set of databases run by librarians. It must be viewed as an intelligence effort in which the right information and data are structured in an insightful manner so as to be valuable and accessible to employees. The job of knowledge management is not bulk – it is synthesis, structuring and insight.

While employee-centric portals can deliver rich and relevant information once it is created, a new breed of knowledge management tools, such as those offered by TheBrain, are emerging to help users make the most of their portals. Tools from TheBrain provide a visual network interface and single point of access to an organization's entire corporate knowledge base. Using an intuitive point and click interface, employees can navigate visually to search the portal for the information they need. The company's technologies are media for organizing and sharing information using artificial intuition.

TheBrain provides immediate access to content regardless of its source. 'Brains' are created by entering 'thoughts', each of which can contain information, files, and links to Internet addresses, other thoughts, and even other brains. Thoughts are organized into as many categories as are desired, and categories can be linked together.

The power of knowledge lies in the insight it provides, and the best decisions and recommendations are based upon concrete facts rather than guesswork. Powerful knowledge management tools allow information to be organized in both vertical and horizontal relational hierarchies that help users establish lateral linkages and relationships between the various pieces of data. Such organization exposes insights and enables the quick retrieval of the right information, enabling companies to share the best learning, which, in the past, may have been buried in vast corporate knowledge bases. Providing this insight is the starting point for empowered employees.

A culture of continuous innovation

Many companies bemoan a lack of creativity and innovation among their employees, but those same employees are often highly creative outside of their work environments. The challenge is to determine why their creativity may be inhibited in the work environment, and determine how to build a culture of continuous improvement, creativity and innovation. Fostering such an environment can be critical to ensuring organizational flexibility, adaptability and, in some cases, survival.

While many companies are founded on an innovative idea, few know how to promote an environment of continuous innovation. Truly innovative organizations are driven by conscious, purposeful, systematic searches for new opportunities. Although there is no one framework for delivering continuous innovation, successful organizations become standard-setters, determining the direction of a new technology or industry. These organizations develop a cultural foundation that leads to new ideas, products and ways of doing business.

The following framework outlines seven interdependent components that we believe are critical for achieving a culture of continuous improvement (see Figure 8.1). Like any other framework, it should be

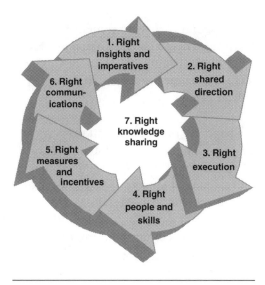

Figure 8.1 Continuous innovation framework.

tailored to an organization's specific needs, using deliberate and systematic planning with both internal and external focuses.

- *Right insights and imperatives*
  Achieving the right insights and imperatives requires a balanced combination of planning approaches. The closed-loop feedback system of the strategic planning approach provides timely adjustment opportunities and ensures the fast delivery of business results. A structured approach to innovation brings together the right people in the right environments to focus on market and customer innovation as well as product innovation. Trigger point planning ensures the earliest possible notification of specific concerns by continuously surveying the internal and external business situation for hypothesized 'trigger points' which initiate predetermined scenario contingency plans. Together, these approaches provide organizations with the foundations to appropriately shape their business strategies.
- *Right shared direction*
  Having the right shared direction requires a dynamic, continuous approach to monitoring and validation of goals and strategies. Regular business assessments serve to continually challenge business performance and identify clear market signals. They ensure the required alignment of strategies and approaches between all parties.

- *Right execution*
  Staying ahead of the competition is as dependent on flawless execution as it is on 'the bright idea'. The right execution is significantly boosted by appropriate capability deployment: the practice of systematically maximizing the value of distinct skills. Process excellence principles are also employed by leading organizations to ensure that right execution begins long before the project starts and continues long after it has finished.
- *Right people and skills*
  To achieve a culture of innovation and creativity, organizations must seek employees who possess not only the immediately required skill sets, but who are also willing to step out of the bounds of their responsibilities and take on new challenges in order to grow with their organization. The personal characteristics of curiosity and drive are key success factors for employees working in a continuous innovation environment. It is also critical that employees are in alignment with the company goals and the strategies for achieving them. This provides the fundamental basis for success.
- *Right measures and incentives*
  The right performance measures and incentives are used to provide an effective management framework, facilitate communications, guide behaviours, foster improvement and assess competitive positioning and operational capability. Effective performance measurement programmes – incorporating a performance measurement framework, the metrics that populate it and the action plans that support it – include baseline assessments and mechanisms to demonstrate the financial impact of performance changes. These programmes help operationalize the organization's vision in everyday, realistic terms and are used to align behaviours and performance from all levels in the organization. Creating mutually supportive groups with a diversity of perspectives and backgrounds, and providing encouragement of innovative work endeavours, goes a long way in promoting a continuous improvement environment.
- *Right communications*
  Right communications requires the effective sharing of information. It refers to the collection, screening and dissemination of the right information to the right people at the right time, and the capability of reaching all levels of employees. Employee portals and broadcast, conferencing and collaboration technologies are enabling organizations to achieve the right communications.

CASE FILE

Introduced in 1987 by Motorola, the Six Sigma management approach is a structured improvement and innovation methodology that has been adopted by a number of leading organizations. This approach strives to achieve business excellence through the application of statistical tools that break down customer process or product requirements into tasks or steps, pinpoint quality problems and determine optimum specifications for each component based upon how they interact. Six Sigma takes nothing for granted and examines even those elements that appear intuitively obvious. It relies on scenarios to analyse the impact of making even the most innocuous change.

Motorola and others, including Honeywell International and GE, have credited Six Sigma with squeezing billions of dollars of cost from their products and processes without sacrificing quality or service. Their successful use of the methodology is largely due to a focus on total employee involvement, training and standardization of work practices and communications.

Motorola includes monetary performance incentives for participating employees and they receive methodology and statistical training to incorporate the practices into their jobs. By 1992, the company was spending $110 million per year on training. Savings credited to Six Sigma include $250 million on failure costs just one year after introduction, and by 1992 an 80 per cent reduction in the cost of quality per unit shipped, yielding a total saving of nearly $4 billion in five years. Six Sigma is now integrated into all Motorola departments.

GE has applied Six Sigma to almost every aspect of its business, including designing products, interacting with customers and fine-tuning delivery. Recognizing that quality requires assuming the customer's perspective, all departments, from accounting to production to customer service, analyse the impact of their internal interaction on a customer's total experience with GE. The company provides opportunities and incentives for employees to focus their talents and energies on satisfying customers. Six Sigma has allowed GE not only to improve the quality of its products and services, but not also to change the way it does business.

● *Right knowledge sharing*
Having the right communication mechanisms can be of little or no benefit if the information delivered is excessive or fails to meet employee needs. The right knowledge sharing requires gathering the most relevant information and synthesizing and structuring it to support work efforts. A range of knowledge management tools exist to help expose the lateral linkages between information and the insight necessary for continuous innovation. They facilitate the speedy

retrieval of information by providing a direct link into the most appropriate best practice materials and information sources.

Motorola, Dell, Dow, Mars, GE, and SmithKlineBeecham have all faced a series of challenges in their attempts to institute continuous innovation programmes. Motorola was hampered by customers unhappy with 50-day cycle times and 40 per cent service levels. Dell faced slow and inaccurate internal reporting of financial, sales and inventory projections, forcing it to forego a new line of laptop computers that failed to meet consumer demand. Dow realized it was behind the times in programme management, information sharing and administration efficiency. GE faced the unenviable task of emulating Motorola's lead and implementing a Six Sigma initiative aimed at reducing the 35 000 defects per million operations – the average for most companies – to five defects per million in every element, in every process. Each organization used a variety of practices from the continuous innovation framework to achieve their goals (see Table 8.2).

Project planning and resource allocation processes

Old ways of evaluating the success of projects according to rigid timeframes and lengthy data collection processes are not sufficient in today's fast-moving business environment. Organizations are now able to share data in its rawest forms, making it available to anyone on an as-needed basis and decentralizing the role of analysis. Project reports and financial statements can be checked on a daily or even hourly basis rather than monthly or quarterly, enabling nimble decision-making and opportunities to perform course corrections mid-stream.

New project management tools are making it easier to plan projects and ensure the effective allocation of resources. These eBased systems allow good ideas from anywhere in a company to be centrally logged and projects formally introduced. Monitoring all projects through one central system also enables the dynamic reallocation of resources to support changing needs and priorities. Three principles can help companies achieve greater value from their project planning process.

First, building slack into the schedule yields a system that is generally more productive than one that operates at capacity. Such a system allows time for experimentation and equips the organization to react to the unexpected.

Table 8.2 Examples of use of the continuous innovation framework.

Organization	Challenge	Leading edge CI practice						CI Outcome
		Right insights and imperatives	Right shared direction	Right execution	Right people and skills	Right performance measures and incentives	Right communications	
Motorola	Five day 'dock to stock' cycle time	✓ Internal		✓ Benefit delivery			✓ Internal	• Cycle times (92%) • Service levels (13%) • Pre-tax profit (24%)
	Reduced customer cycle time to less than one day	✓ Internal	✓ CI planning process	✓ Benefit delivery		✓ Metrics		• Cycle time goal 3.5 days • Process steps: 17 to 10 • Elapsed time/Value time ratio: 23.3 to < 5 • Department able to handle 20% workload
	Improved parts distribution process for internal clients	✓ Internal	✓	✓	✓ Training and development	✓ Metrics		• Cycle time (95%) • Inventory turns: 4 to 12 • Saved $1.6 million by number of purchase orders written • Inventory space (75%)
	Improved production cycle time	✓ Internal	✓ Target setting	✓ Benefit delivery				• Cycle time 64% in 5 months • Floor inventories by $69 000 • Saved $46 000 taking one day out of cycle time
Dell	Improved ability to share information internally		✓ Target setting	✓ Benefit delivery	✓ Behaviour	✓	✓ Internal	• Sales, profits and inventories available 5–10 days after close of quarter compared to 30 days previously • Set goal of 2–3 days
	Closer connection with corporate and retail customer needs	✓ External			✓ Behaviour	✓ Performance management	✓ External	• System that directly links customers and vendors to manufacturing, finance, service, products, sales, where feedback can be turned into incremental improvements
Dow Chemical	Programme based on customer-focused improvements		✓ CI planning process	✓ Benefit delivery	✓ Training development		✓ External	• Productivity • Plant safety awards • CI production plant programme regarded as best practice, copied by Ford and GM

Table 8.2 (continued)

Organization	Challenge	Leading edge CI practice						CI Outcome
		Right insights and imperatives	Right shared direction	Right execution	Right people and skills	Right performance measures and incentives	Right communications	
Roche	Increased effectiveness in developing and managing human resources	✓		✓ Change management and programme management	✓ Training development and behaviour			• Operating unit HR plans and value-added corporate services • Functional barriers within HR function, resulting in flexibility • Streamlined HR processes • Business knowledge and operating area experience among HR staff • Individual and team competence
General Electric	Reduced defects from 35 000 per million operations to 5 defects per million in every element, and every process	✓ Internal	✓ CI planning process	✓ Benefit delivery	✓ Behaviour			• Operating margin – from 14.8 to 15.7 in 1 year • Six Sigma responsible for $700 million in corporate benefits in 1997, 1.2 billion in 1998 • GE appliances completed 1000 Six Sigma projects in 1997 with savings of $43 million and goals of 3000 and $100 million respectively for 1998

Second, using tools and portals to gain instantaneous access to information enables analysis and reallocation of resources at the time that issues arise.

And third, creating pools of people to respond to the unexpected ensures that a project will not become derailed when unforeseen events occur.

Conclusion

eWorking is the way of the future – it makes good business sense. The value of employing new tools such as employee-centric portals, standardization of processes, new approaches to knowledge management and a commitment to continuous innovation can be quantified and valued, with companies like Cisco, Microsoft and Oracle counting dollar savings in the tens and hundreds of millions. The savings are generated via the reduced need for personnel through the streamlining and automating of processes and systems and the efficient exchange of information. Employees are able to redirect their time away from non-value-added activities, such as the arduous task of gathering data, to focus on the actual analysis.

Companies now have the tools to create a new breed of knowledge workers that thrive in a culture of innovation. The companies that succeed in the eEconomy will be those that look beyond the short-term drive for success and focus on building innovative organizations that draw talented individuals.

9 Looking over the cyberhorizon

Introduction

In this book we have explored a range of waves of change, mostly driven by technology, that are having an impact on the ways in which companies design and manage their supply chains. The intensity of these waves is high, posing significant challenges for business. However, even as business leaders respond to today's challenges, they must also keep an eye on the horizon for the first signs of the waves to come. As someone put it: 'The future is already here, it is just unevenly spread!'.

It does not matter that many forecasts will be proven wrong – the important thing is to look ahead with anticipation rather than anxiety. Anticipation will help transform guesses into self-fulfilling prophecies as we work to identify and shape the issues before us. Despite the fact that many forecasted changes will fail to materialize, companies will need high-quality systems to monitor new developments in their own industry as well as in complementary industries that will have an impact on their supply chains. Beyond such monitoring, those companies that can influence the future of their industries will surely dominate.

In this chapter, we explore three final questions:

- What are some of the future trends that we expect to have the most impact on supply chain management?
- How can the eight cultures of value chain competitiveness help to deliver success in this future environment?
- Why will inspirational leadership play a critical role in enabling organizations to deliver required changes to supply chain management?

Nine trends for the future

Many people and organizations specialize in predicting the future. We have considered their work while envisioning the future of supply chains, and identified nine key trends that we believe will continue to transform supply chain management. In some cases, these trends make the supply chain more complex, in others, refreshingly simpler.

- A new landscape of companies will be driven by a new multinational focus on core capabilities.
- New types of supply chain service organizations will emerge.
- Broadband technology will enable the next stage of the Internet revolution.
- Mobile commerce (mCommerce) will expand business options – and complexity
- Interactive capabilities will become increasingly important in managing transactions.
- Cities will require new distribution services.
- Intelligent city utilities and associated services will emerge.
- The roles of government and regulators will be redefined.
- Supply chains will grow increasingly vulnerable to shocks and security issues.

A new landscape of companies

In earlier chapters we discussed the need for traditional bricks-and-mortar multinationals to reinvent themselves to embrace the age of the Internet. Change in these companies will be driven by the growing visibility of the best getting better and the rest getting worse. As traditional companies realize the implications of the changes required to remain competitive, we will see a new wave of outsourcing and alliances that will fundamentally change the nature and composition of multinational companies.

We believe there will be a rapid and sustained move towards the use of specialized, back-office shared services operating companies – once these companies prove that they can provide improved service at lower cost. We also anticipate the development of specialized contract manufacturing operations and service companies serving multiple sites, and a growing separation between manufacturing and sales and marketing as companies focus on their core capabilities. Similar to Nike's operating model, many of these companies will concentrate purely on sales, marketing and product development operations, leaving the manufacturing to outside service providers.

This shift to core capabilities creates a tremendous opportunity for contract manufacturers, but we also expect to see the emergence of companies that offer other specialized services such as supply chain optimization operations to manage the synchronization of many players in a value chain, or ownership and operation of assets such as storage and transportation. We use the term Fourth-Party Logistics™ (4PL) to describe this new breed of supply chain service provider.

The net effect of these changes will be a new landscape of companies. The division of multinationals into functions that are managed internally and externally will lead to greater visibility about where value is created in their supply chains. Few multinationals are expected to sustain their current traditional structures over the next five years.

New supply chain service organizations

The trend toward new types of supply chain service providers is likely to accelerate and be dominated by two types of company: those that specialize in owning and operating core assets and those that specialize in supply chain optimization services.

We expect the supply chain asset companies will focus on consolidating assets that serve multiple companies across industries. For example, cross-company operations are likely to be consolidated in The Netherlands in the same way that multinationals have already consolidated their northern European logistics in that country. The supply chain asset companies, which are more advanced in the USA, emerging in Europe and increasingly present in Asia, will be playing a long-term, low-margin game. Their competence will be in rationalizing assets and in providing low-cost service operations. Asset operations will become more global once the earlier stages of regional rationalization are achieved.

Supply chain companies that specialize in supply chain optimization are still in the early stages of development. They have their roots in companies such as Caterpillar Logistics Services, which have developed their expertise by applying advanced information technology to optimization problems such as effective 24-hour delivery of slow-moving parts on a global basis. A new type of smarter supply chain operator will be driving these companies. The supply chain optimization specialists will take responsibility for managing complex cross-company and cross-industry optimization services, collecting data from multiple organizations, and analysing and optimizing operations on a real-time basis. They are likely to be staffed by people drawn from a combination of existing logistics companies and internal functions, mathematics or statistics graduates and supply chain management consultants. These service providers will make extensive use of advanced technologies and sophisticated optimization modelling techniques. Their value proposition is their ability to offer clients access to advanced solutions without the cost of hiring and retaining the people and software required to perform the function internally.

We see these companies operating command centres that are not unlike those of a military operations room. Their challenge will be to optimize complex supply chains, increasingly in real time, to help companies synchronize operations.

Broadband technology

Broadband technologies provide the potential for Internet connection speeds that are hundreds of times faster than today's commonly used dial-up modems. Such speeds will enable continuous online access for consumers from anywhere in the world, at a competitive rate, with rapid response times and a multitude of network benefits.

The implications of broadband technologies for the supply chain are significant. Huge amounts of data can be passed much faster between ever-increasing numbers of organizations and people on a global basis. This type of technology will remove the constraints of access and time from participants in the supply chain. It will also raise expectations of a company's ability to operate 24 hours a day, 7 days a week and 365 days a year. Round-the-clock operations can be achieved using multiple locations around the world.

Broadband technologies will create new possibilities for monitoring and tracking devices for everything from plant process control to smart appliances and product tracking, and will lead to new ranges of product and service offerings. For example, collaboration between retailers and central suppliers can be improved at an industry level through an exchange of real-time point-of-service data, improving the flow and location of finished goods. The idea of 'sell-one, replace-one' (a system where point-of-sale information is available to the point where replacement of an item occurs only as a direct result of a sale of that item) will come that much closer to being a reality. However, with these benefits comes the challenge of data management and analysis, similar to that faced by supermarkets when they first introduced checkout scanners. The first obstacle is to get the process working, the second is to understand the data and how it can be used to optimize the supply chain.

CASE FILE

Broadband Technologies

The five major broadband technologies that compete for the most efficient delivery of electronic data are digital subscriber lines (DSL), hybrid fibre coax, fibre optic cable, local multipoint distribution services (LMDS) and satellites.

DSL relies on existing copper telephone wires, with frequencies higher than those carrying normal telephone conversations. Hybrid fibre coax utilizes the existing combination of fibre optic lines and coaxial cable, and provides television's existing infrastructure. Fibre optic cable can handle large volumes of different types of communication signals simultaneously, including telephone, television, videoconference, movies, telecommuting and Internet traffic, with some providers laying fibre optic alongside original coaxial cable. LMDS is a wireless system that delivers data using high-frequency radio signals through rooftop antennae. Satellites operate by receiving and re-sending radio transmissions from ground-based antennae.

DSL is 50 times faster than the standard modem, while hybrid fibre coax is 200 times faster. Although fibre optics provide the fastest possible broadband connections, it is prohibitively expensive in those areas without cables. LMDS, on the other hand, can be deployed quickly and inexpensively. The most effective broadband capability is, however, satellite technology. Low orbiting satellites deliver broadband interactive services at twice the speed of LMDS, three to six times faster than cable modems, and up to 12 times faster than DSL. The cost of installing these satellites is high, but as global use of the Internet and other services increases, demand will likely justify the costs.

mCommerce

Developments in mobile commerce (mCommerce) will create a whole new range of supply chain options – and increase complexity. Currently, mobile technologies are effective for voice transmission but weak for data. Early attempts at mobile Web access have also shown that most

Web sites are not prepared to send the type of data that is suited to a mobile device. As a result, most smart mobile phones, personal digital assistants (PDAs) and wireless application protocol (WAP) devices currently provide an unimpressive Web service. However, this will change rapidly.

In the future, wireless mobile technology will provide a personal, omnipresent personal channel offering a range of information, interactive communications and customized services. Mobile telephony will deliver pictures, graphics, video communications and other wideband information, allowing users instant access to information. It will change the dynamics of supply chain management, with greater use of mCommerce technology in process control, tracking and tracing, and security applications, leading to greater visibility of supply chain activities to a wide range of participants. mCommerce will allow retailers and manufacturers to learn more about consumer buying patterns, leading to a more efficient flow of product through the supply chain and a more efficient alignment of the corresponding supply chain infrastructure – and growing demand for mass customization. Automated databases will allow companies to access huge numbers of consumers to deliver a wide range of value-added services.

Wireless technology will not be limited to mobile access – almost any appliance can become 'smart'. The potential for a networked home is tremendous. There will be a proliferation of intelligent home appliances, such as refrigerators that monitor their own mechanical condition and call for repair *before* they fail, or local supermarkets that check the contents of customers' refrigerators to compile shopping lists as supplies diminish. In a world of smart appliances, the line between product and service will blur. And, if the service becomes sufficiently valuable and lucrative, service providers will even subsidize the purchase of smart appliances just as mobile phone companies do today.

Third-party service providers have recognized the potential of combining wireless mobile phone technology with payment services. Pilot schemes around the world allow consumers to 're-load' their mobile phones for more talk time, pay for unattended parking, buy soft drinks from vending machines, and pay at car washes. Goods and services are charged through the telecommunications operators and show up on customer bills.

CASE FILE

Bluetooth

Companies from the mobile telecommunications industry are placing their bets on Bluetooth wireless networking technology. Formed by IBM, Intel, Toshiba, Ericsson and Nokia in 1998, Bluetooth is used to build personal area networks that connect mobile phones to handheld computers and tiny printers in the immediate vicinity of an individual. Using a globally available frequency range, as many as eight devices can be connected to one *ad hoc* network with a range of about 10 metres. So the laptop in your bag could use the wireless phone on your belt for Internet connectivity, or the phone could pull a number from the handheld in your pocket.

Interactive capabilities

In addition to improved bandwidth and mCommerce technologies, we expect an increase in interactions between buyers and suppliers – and they are unlikely to be limited to transactions. As real-time communications tools become more widespread, many more interactions will be linked to decisions and feedback. Supply chain organizations will need to be prepared to manage a proliferation of communications between customers and suppliers.

Early forms of interaction will no doubt be driven by Interactive TV (iTV). For the first time, consumers will experience the increased value of interacting and transacting through a familiar device. Imagine banks advertising on television their latest service while simultaneously providing a channel link to sign up for it, or avid soccer fans ordering a pizza using an onscreen link to their local pizza outlet during the soccer game. Pharmaceutical companies could evaluate the efficacy of a new pain reliever by asking viewers to complete an on-screen feedback form during an advertisement for that product. iTV's capacity to target customers in real time provides manufacturers new opportunities to test product designs by providing links to a three-dimensional laboratory.

Interactivity will touch upon a wide range of services, particularly where there is existing dissatisfaction about the quality of service or effectiveness of operations. Pharmaceuticals and medical care are great examples of industries in which significant amounts of money are spent on drugs, but there are very few real statistics on how effective the products are. We see an increasing demand for interactive services around a wide range of goods and services, and a strong demand for flexible supply chain response to feedback.

CASE FILE

Interactive television (iTV)

iTV is developing at an extraordinary rate. In early 2000, there were some 7 million such devices in Europe. By 2006, the Internet may be accessed more often through iTV devices than through personal computers, as almost every home in major industrial countries will own a TV connected to an interactive service through a set-top box. Research shows that the most promising feature of iTV is its ability to record and store programmes that can be replayed on demand. However, other information services, such as teletext, which allows scanning of local news and information, programme and content downloads, programming with embedded content such as searchable statistics during a sporting match, intelligent programme guides and interactive shopping channels, are likely to become popular features.

With the potential boom in television equipment replacement, hardware manufacturers will have the opportunity to move into the home server and storage arena. Advertising will need to be more compelling and tailored to the interests of segmented viewers. Given iTV's capability for gathering data about its viewers, downloading a movie could, for example, trigger a series of ads based on what the viewer has watched and purchased in the past.

Distribution services in cities

The need to respond to the economics of home delivery and increasingly demanding consumers will drive the need for new types of distribution services in cities. Already, most cities cannot cope with their traffic levels, and access to on-demand goods and services through the Internet could lead to more home delivery vans on city streets. This could be a positive or negative change depending on the net effect on consumer behaviour – Internet versus traditional. There is the potential for a fundamental realignment of the ways in which goods and services are distributed to reduce traffic, congestion and pollution.

We expect dual usage of localized distribution facilities to make Internet transactions manageable and cost-effective. The role of local facilities in selling and taking orders will diminish as standard commodity goods such as groceries, staples, appliances and alcohol are purchased online. The challenge will be to close the loop between buying online and providing cost-effective fulfillment. Therefore, we expect to see the emergence of mini-warehouses or cross-docks in local areas to meet the challenge of home delivery. Picking and packing may be done in vast centralized facilities – Webvan has developed such a model.

CASE FILE

Interactive HealthCare

The networked health system will ensure secure access to online medical data for both patients and health professionals. Medical histories will be accessible online or stored on a smartcard. Diagnosis, treatment and prognoses will be recorded and stored in the patient's medical history file. For specialist consultations or second opinions, the time saved in history taking will be enormous. Patients, familiar with the technology available to them, will investigate treatments for their own ailments, including complementary therapies, using Internet searches that retrieve information consistent with the symptoms identified by the patient. The potential for a more proactive, informed approach means higher patient satisfaction with the services provided, reducing the number of complaints to government health bodies. The greater efficiency afforded by this online approach to consultations means that medical professionals have more time to treat a larger number of patients with more complex symptoms, and invest time in keeping up to date with the latest breakthroughs in medicine.

A system of networked databases for medical information also ensures a reduced number of 'doctor shoppers'. Doctor shoppers abuse the medical system by visiting large numbers of doctors and obtaining multiple prescriptions. As any prescribed medications will be recorded on the patient's medical history file, either online or by smartcard, doctors will be able to detect these doctor shoppers and improve overall patient care by offering appropriate counselling, identified by checking the portal for affiliated counselling services. This saves on doctors' time and taxpayers' money. Thus the network provides a direct channel for the delivery of related healthcare products and services, ensuring one-to-one relationships with patients.

For those patients with identified illnesses, the effects of the new approach to healthcare would be enormous. Online systems can provide a situation-specific portal for healthcare. Combining sensors, the Internet and embedded computers, they create an intelligent appliance that continually monitors the needs of these patients and responds with appropriate individualized services. Integrating technologies such as smartcards and voice recognition, this appliance displays a list of personalized health reminders, such as doctors' appointments and medication alerts. The patient is able to schedule appointments by linking to the surgery's online calendar, which lists available consultation times. Once the appointment is selected, it is automatically updated in the patient's PDA. On receiving pathology report results online, doctors are able to prescribe and order the necessary medications, notifying the patient by email. Using public key infrastucture (PKI) or smartcard details to pay for the transaction, the patient nominates a preferred delivery time and location. At the same time, health insurance companies are notified of the medications purchased, and rebates are automatically transferred to the patient's bank account.

The online system's ability to monitor drug dispensing has many benefits for both the patient and the health industry. For both hospitals and home medicine cabinets, orders can be placed when stocks run low. Technology keeps track of supply, and reduces the likelihood of incorrect dosages, often attributable to human error.

However, we do not believe that home delivery from these large facilities to homes will be economically practical. Instead, we expect a number of existing local operations to take on a multi-role. These operations are likely to be facilities such as petrol stations, public houses and bars and local post offices. As these facilities are redesigned and re-roled they will be increasingly able to satisfy the 'last mile'. This can be achieved either by consumers collecting goods or by highly localized delivery services. We expect these new types of distribution services to develop rapidly, either through increasing consumer purchases via the Internet or because of traffic management systems that charge entry and exit to local areas.

The intelligent city

Cities are coming to increasingly dominate world commerce and wealth. Cities like New York, Chicago, London, Singapore and Frankfurt, and geographic basins like Silicon Valley, are capturing a growing share of wealth and attracting the best talent. The value of transactions in particular cities is huge and most often underestimated. For instance, the value of transactions in Milan or London exceeds all Italian or UK exports respectively. From this we expect to see an Internet-driven move towards intelligent city utilities (iCity) as public and private bodies combine to create local transaction hubs.

Recent observations have shown that as the number of online consumers in one location reaches over 40 per cent, a localization effect occurs. An increasing range of transactions will be conducted over the Web – for example, booking a tennis court, paying local taxes or booking a baby-sitter. Some industry experts reason that if enough local content and services are available online, and if Internet access is low-priced and widely accessible, up to 70 per cent of Internet traffic would take place within local communities. Companies will need to consider increasing their local content, not just translating it for the Web, but localizing the content so that it looks and feels like it was created in that city.

This means that the physical expansion of the city may not be a requisite for economic growth as the patterns of city living begin to change. Schools could provide online homework, taking advantage of innovative approaches to education, and become community resources after hours. Internally, public sector services could be improved through the provision of online transactions for taxes and benefits, as well as information and services, reducing the financial and administrative burdens of government agencies. This would also serve to stimulate and support community eBusiness activities.

Citizens of the iCity will have greater information and greater choice, with opportunities to acquire inexpensive home computing appliances and high-speed access to government, local businesses and education. In addition to hardware, citizens will receive a bundle of Internet-related services, such as email, Internet access and Web space, adding to the viability of the work-from-home option. In this way, they benefit as customers and voters.

CASE FILE

Traffic management in the iCity

Although various technologies are already available for the management of traffic and public transport, the iCity's ability to coordinate these and other technologies using the broadband access network means a fully integrated approach to the effective control of transportation. Policy choices about levels of traffic flow, impacts on health and the environment, and the desired level of public transportation services will inform this system. Not only will the iCity's capacity for optimal traffic management allow greater control over traffic flow, but it will facilitate a more effective, intelligent response to unpredictable events, such as major accidents.

The iCity's intelligent traffic management system will help to save lives by reducing response times to emergency situations. General safety will be improved as automatic guidance systems will ensure optimized spacing of moving cars, which have their engine RPMs, antilock brake systems, as well as air conditioning and tyre pressure, controlled by software. Although initial investments may be high, the integrated approach to traffic and transport control will reduce running costs in terms of road maintenance, and time and fuel prices for public services. The fact that transport will be more effectively managed should facilitate greater use of public transport, as people will have reliable, real-time data about travel timetables on their PDAs or other Internet appliances.

The information collected by traffic management technology would allow the iCity to match road haulage companies to customers, reducing congestion by removing empty lorries from the roads. A quarter of freight lorries that make use of European trade routes today are empty, and many are travelling well below capacity limits. As key business stakeholders in the iCity, manufacturers, hauliers, warehouse operators and distributors could list details of their requirements on a nominated freight-trading hub, allowing hauliers to bid for contracts. This open forum for exchange would circumvent the problem of lorry fleet operators trying to find goods to carry back on return trips. Several interlinked hubs would provide entry for other businesses in the transport industry to capitalize, for example, on improved engine combustion techniques and low sulphur fuels coming onto the market. This auction service would also reduce congestion and pollution substantially.

The implication for the supply chain of intelligent cities is significant. The consolidation of transactions within concentrated localities offers opportunities to build new types of supply chain services (such as eFulfillment, local distribution services and the reinvention of postal services) within an individual city. It also has the potential to increase supply chain complexity. Companies will have to drive out a degree of commonality and standardization between large numbers of intelligent city utilities. Intelligent cities will increasingly also be challenged to find cost-effective means of servicing rural communities by national governments.

The role of government and regulators

The Internet, and the subsequent changes to the ways in which transactions are managed, has huge implications for the government and regulators. We include this section because there are interesting and valuable supply chain opportunities associated with eGovernment transformation and regulatory changes.

For government, there will be pressure to change internally and support the development of new capabilities externally. Examples of the major issues governments will face are:

● Reinventing supply chain processes within national and local governments, such as procurement processes, elimination of redundant processes, online payment of taxes, provision of supply chain information, retraining of people and outsourcing of non-core activities.
● Redefining the role of postal services
● Taxing eCommerce transactions.
● Responding to the challenge of intelligent cities versus rural communities.
● Educating young and older people in eCommerce-related skills.
● Managing pollution and traffic (public vs. private).
● Reusing and selling government knowledge assets such as databases, maps and public information resources.
● Rethinking schools from being single-use physical assets to virtual and physical community assets.

These issues will seriously challenge governments at a time when their ability to manage markets will be seriously diminished. In the same way that we expect traditional multinationals to disaggregate and reaggregate in new forms, there is potential for eGovernment transformation. The challenge for commercial supply chain companies is to reposition to help eGovernment achieve this change.

The development of new technology drives up the demand for highly skilled workers. In most societies, the school system is still designed for a more industrial society. This type of school system tends to produce a ratio between manually and professionally skilled labour of about 80:20. In an economy with strong demand for knowledge workers, such an approach will create social exclusion or a digital divide between those who can and cannot use Internet technologies. A key challenge for government will be to manage the digital divide so that it does not become divisive in societies that will be increasingly difficult to manage centrally.

For regulators, the increasing pace of change will place requirements for new and more complex decisions by policy makers on new business models. This will require new capabilities to manage policy decisions at a national and supra-national level to decide regulatory policy in areas such as:

- eMarketplaces and what constitute competitive and non-competitive practices.
- Cross-border trading and taxation.
- Ownership of products and assets.
- New business models and forms of operating companies.
- Movement of work to low-wage economies.
- Responsibilities for serving urban and rural communities.
- Management of round-the-clock services.

Our experience is that eCommerce and new business models have generally left regulatory regimes behind. Regulatory responses have often been 'wait and see' or 'we will get to it in the future'. Moving forward, it will be difficult for regulators to exert control in new markets and new business models unless they present a more united and responsive front.

Shocks and security issues

While new technologies will yield tremendous benefits, they will also introduce significant risks to global businesses. A shock to a supply chain system, such as the restriction of petroleum deliveries in Europe through a political protest, can quickly bring a highly efficient supply chain to a standstill. Businesses and governments will have to consider that high levels of automation and efficiency introduce vulnerability and brittleness into supply chains. We can expect to see unexpected crashes in supply chain effectiveness where there is not careful attention to shared redundancy and security.

In its simplest form, the Internet has elevated the virus from a simple nuisance to a potential weapon for cyber-terrorism that could affect any device connected to an extensive network. For example, recent clashes in the Middle East have been accompanied by Web site attacks. Attacks on company Web sites are likely to become more common as pressure groups, individuals with grudges and hackers see the opportunity to make a highly visible impact with relatively little effort.

Security issues are clearly on the minds of potential eMarketplace participants. The very technology that allows many-to-many collaboration through portals must also offer data security for participants. The success of emerging eMarketplaces will require a high level of trust that companies can share their systems and data without fear of attack or compromise. Security capabilities will be a growing area as companies seek to defend their data, knowledge and transaction assets.

Issues of security of transactions and privacy will grow in significance as more consumers and businesses conduct more transactions online. Consumer and business fears of privacy invasion, if not addressed, will present a barrier to online success. We can expect to see increasing resistance in some communities to the visibility that the Internet can bring to their behaviours and supply chains. This will be particularly true where price visibility is heavily eroding margins.

The eight cultures of value chain competitiveness and future success

In the early parts of the book we introduced the eight cultures of value chain competitiveness as a framework for future success in multinational companies. Good management is about balance and a culture of good habits. This is as true today as it has always been. In this section, we review how these eight cultures relate to the projected trends of the future.

Operational excellence and continuous innovation

Operational excellence and continuous innovation will remain the bedrock of a company's competitive advantage. As traditional multinationals disaggregate and reaggregate into new types of company, operational excellence will become more important, as the new companies will have an absolute focus on excellence in their chosen areas. As these new types of company form, weaker performers will be highlighted, and will experience increasing pressure from shareholders to change. Even marginal players in a value chain will be exposed if they cannot connect and manage their data disciplines. Overall, we expect significant improvements in operational performance and a reduction in variability of performance between companies and between business and government.

Extending into customers and suppliers using Internet technologies

In the future, efficient connectivity between customers and suppliers will be the norm, not the exception. Supply chain collaboration will simply be a part of the way that companies do business. Those companies that have limited plans to extend their reach through Internet technologies will be challenged if they cannot collaborate with their business partners. The future will be about a connected world, where people will only want to do business with those companies that know how to connect.

Compressing the supply chain

We expect to see a whole range of activities taking place in the future to eliminate waste of time and resources in the supply chain. Demand and supply planning packages linked into ERP systems are becoming very common. We expect this to radically reduce inventory levels through MRP linkages, available-to-promise capability, collaborative forecasting and vendor-managed-inventory (VMI). As organizations become more experienced with sharing data, ideas, assets and new business opportunities, we expect to see shorter, more integrated supply chains working together in synchronization. New communications technologies and mCommerce will aid this trend. The inclusion of government and

local transactions in these value chains will also make the prize of
supply chain compression even more rewarding.

Creating market-level contingency

It is clear that one of the implications of downsizing in many industries
is that companies are becoming increasingly lean and brittle. We expect
to see companies recognize this and do one of two things. Firstly,
nothing – thereby leaving themselves at the mercy of the gods. These
companies may encounter few problems initially, but at least some will
have significant problems over time. Secondly, we believe we will see a
range of innovative market-level solutions emerging from groups of
companies. At one level, this might be building up industry-level
reserves of inventory and excess capacity in manufacturing assets. At
another level, it is conceivable that eMarketplaces and Fourth-Party
Logistics™ providers may take on responsibility for managing
contingency within an industry or functional area. What we can be sure
about is that there will be failures among those that spread themselves
too thinly and do not plan for unexpected shocks.

Optimizing price and revenue

Much of the volatility in supply chains is caused by human decision-
making. We expect greater visibility of supply chain data and decision-
making to smooth out some of this volatility, particularly if there are
supply chain optimization services helping with the decision-making
process. There is also the opportunity for the much greater use of
dynamic pricing and revenue optimization techniques to be used to
choke off excess demand or encourage changes in supply. This has
always been a bit of a black art in the past. In the future, we expect
greater visibility of supply chains and data to encourage greater use of
pricing to regulate volatility in supply chains. Companies like Dell
Computer are already experimenting in this area – expect more to take
advantage of this lever.

Learning to operate in eMarketplaces

The growth in the number of B2B eMarketplaces will decrease; however,
the volume of trade will increase for those that are successful. We expect
companies to struggle with eMarketplaces for at least the next few years.
Those that are successful will have great assets to take advantage of
particularly if they have developed effective 'onboarding' strategies and

plans. The number of exchanges will be rationalized to the point where there will be a few vertical exchanges for each industry sector, and a number of horizontal exchanges cutting across to provide complementary services. A world characterized by a network of interconnected exchanges will give rise to an entirely new type of supply chain collaboration and synchronization environment.

Significant technology problems still need to be solved – such as integration of technologies between the companies and marketplaces and the complex rationalization of internal systems. The companies that are best positioned to take advantage of eMarketplaces are those that have the foresight to have achieved and sustained operational excellence and can therefore concentrate on dominating new and complex connectivity issues.

Supporting people to change and perform

There is already a war for the best talent in the new economy. This is not about headhunting Internet entrepreneurs – that game is now nearly over. This will be much more about attracting, retaining and empowering groups of talented people who understand how to operate in teams in an Internet-enabled environment. These are people who like to operate without many of the normal constraints of traditional multinationals. These people will go where they feel their talents are best used and rewarded.

The 'Best' multinationals are also realizing that it is not just the quality of people that counts. It is also the level of variability between the performance of the best and worst. We believe that the most common characteristics we see in the 'Best' multinationals are low variability of performance around the average and the space for 'stars' to perform in a team setting. Typically these multinationals also, strangely, have an illogical obsession with some area of process excellence that binds the organization together in a common set of norms, as do Bechtel in project management and Cisco Systems for the 'Cisco way of business'. In the future, we expect to see new ways of managing people in distributed and global environments that retain the good habits of the past and enable more people to contribute at a higher level of performance. The biggest challenge for traditional multinationals is being brave enough to let people operate in this way across value chains.

Building new business models and relationships

In the past, we were constrained by the business models that we could work within. Indeed, for most of the 1970s and 1980s, logistics management seemed to be preoccupied with 'integration' through various means – mergers, acquisitions, organizational changes. For the most part these initiatives were ineffective, if not within organizations, certainly *between* organizations. New technology has provided the capability to enter a virtual world where everyone can still have access to the information, accessed from anywhere. This has opened up many options for new business models.

We expect to see a whole new range of supply chain and other related business models. Many are already under development. Many will fail – because of execution rather than the quality of the idea. We expect to see changes in the way we manage supply chains with smaller more nimble companies constantly innovating to take advantage of new business opportunities.

Emerging age of inspirational leadership

The evidence from stock exchanges around the world indicates that those organizations that are able to produce superior financial performance on a sustained basis have been able to achieve 'alignment' between their customers' requirements, their corresponding response to these requirements, the cultural capability of the organization and the leadership of the chief executive and his or her top excecutive team.

The recent changes that have been enabled by Internet technologies have brought with them a new interest in inspirational leadership. Some companies provide very good examples of the power of inspirational leadership – Sun Microsystems, Oracle, Cisco, Intel and GE. The common factor in these high performing enterprises is that they are dominated by inspirational and indeed motivational leaders. These executives are up-front about the fact that they are the leaders and sometimes act more like politically astute entertainers. They are very good at communicating their vision and intent to their organization and other stakeholders.

There is a growing desire among employees to witness their leaders using and embracing leading edge technology and supply chain approaches coupled with a high degree of business savvy. Due to the speed with which the business environment is changing, these leaders are largely operating in a turbulent, uncertain and competitive world where entrepreneurism, creativity, speed of decision-making, a good understanding of their industry and a high tolerance for ambiguity are among the most essential qualities required for survival and success.

A closing thought

The very technology we have discussed in this chapter is making the world ever smaller – but also increasing complexity, both real and perceived. Equally, the same technology, used in the right way by the appropriate people, has the capacity to resolve huge levels of complexity. It is almost certainly true that the world of tomorrow will belong to those who are brave enough to move early and decisively. Inspirational leadership will be an essential ingredient for success. New emerging technologies will strengthen the supply chain in operational ways, at both enterprise and industry levels. We predict an 'Industrial Revolution'-scale change in the management of supply chains as we move into value chain competition. In time, there will be a spillover from enterprise and industry levels to the city state, where anything of a network nature will be affected – and the impact on citizens' lives becomes even more pronounced. In short, supply chain considerations will dominate our lives more than ever before, well into the third millennium.

Glossary

3PL   Third-party logistics; the outsourcing of transport and warehousing to a third party.

4PL   Fourth-Party Logistics™, an Accenture registered trademark now adopted as a generic term for the outsourcing of a company's entire supply chain requirements. The 4PL is in effect a virtual management enterprise with preferably two main equity partners who bring logistics business to be managed, and several minority equity members who bring specialist skills.

Applications software   Software which performs a specific task or range of tasks.

ASP   Applications software provider – a third party which provides and runs applications software over the Internet on behalf of its clients.

Available-to-promise capability   Ability to supply against agreed delivery conditions for goods which may either be in inventory production or planned production.

B2B   Business-to-business.

B2B2C eFulfillment   Integrated **eFulfillment** throughout the entire supply chain.

B2C   Business-to-consumer.

BAMs   Bricks and mortar, i.e. traditional companies (the opposite of **dotComs**).

Best-of-breed   Best company, product, service etc. in its class.

Broadband technology	New generation communications technology capable of handling large numbers of individual channels and/or channels carrying high levels of information, for example video.
Browser-based user interface	Makes application software look and feel like Microsoft Internet Explorer and/or Netscape Navigator.
Business models	Alternative schema for conducting a particular class of business or transaction.
CAD	Computer-aided design.
C-level	Senior Executive Level, for example, CEO, CFO, COO, CIO.
Collaborative forecasting	Forecasting based on a sharing of information between supply chain partners.
Company mini-hubs	Aggregated procurement, fulfillment and support provided by large companies to other suppliers and customers in their value chain.
Connectivity	Linkage between customers and suppliers.
Consolidator	Service provider who receives products from multiple suppliers and consolidates them into single deliveries for individual customers.
Content aggregation	The bringing together of supplier information to build an electronic catalogue.
Content conversion	Standardization of catalogue content to enable rapid searching.
Content rationalization	Process of cleaning up and organizing supplier product data prior to **content aggregation**.
Co-opetition	Collaboration between competitors to meet a specific customer's requirements.

Cost-to-serve   The cost to a supplier of meeting a customer's service level expectations.

CRM   Customer relationship management.

Cyber-terrorism   Disruption of eCommerce activities by hacking, spamming, viruses etc.

Data integration   The bringing together of data from disparate sources.

Data mining   Analysis of generally large volumes of data (for example transaction data) to reveal significant patterns and predict future customer behaviour.

Definite auctions   Alternative term for 'pre-configured supply options' in which buying companies can place options to purchase goods and services on suppliers' Web sites on the basis of price and other variables, and automatically trigger bulk purchase of an optimized quantity or a change of supplier when its conditions are met.

Demand chain synchronization   Optimization of logistics operations and infrastructure from a market perspective involving sharing of forecast and capacity information and industry level management of assets.

Development portal   Single point of access to all tools, information and resources required for product development, open to all parties, internal and external, involved in that project.

Digital connectivity   Linkage between customers and suppliers over the Internet.

Digital divide   Division in society between those with and those without access to Internet technologies.

Digital transaction hub   Means of integrating groups of companies who routinely buy from and sell to one another and therefore do not require the full facilities of an eMarketplace.

Disinter-mediation	The removal of intermediate stages such as distributors and stockists in a distribution channel.
Domain expertise	Expert input required to rationalize information in a specific field or domain into technically robust and user-friendly formats during **content rationalization**.
dotBAMs	Bricks and mortar companies (**BAMs**) which have built on the experience of **dotComs** and **dotCAMs** to develop an eCommerce capability.
dotBOMs	**dotComs** which have bombed.
dotCAMs	Clicks and mortar companies which have combined an effective Web presence with an effective, responsive supply chain.
dotComs	Classic Internet **start-ups** characterized by an ability to manage relationships and an inability to manage supply chains.
DP	Demand planning.
Drop-shipping	Customer delivery via distributors' regional warehouses.
DRP	Distribution resources planning.
DSL	Digital subscriber line.
Dynamic pricing	Managing prices in real time in response to changing supply and demand.
EAI	Enterprise applications integration – common framework for integrating end-to-end business processes and data.
eAuctioning	**eProcurement** technique providing for competing bids for goods and services offered over the Internet.
eBusiness	Any and all business conducted over the Internet; cf. eCommerce.

eBuying	Automated order placement over the Internet, also referred to as **eRequisitioning**.
eContracting	Identification of and contracting with sources of supply directly over the Web.
eCRM	Electronic customer relationship management.
eDesign	Internet- and intranet-enabled integrated and collaborative product design.
EDI	Electronic data interchange.
eFulfillment	The delivery to business or consumer customers of products and services ordered/purchased over the Internet.
eGovernment	The response to and exploitation of Internet technologies by local and national government.
eIntelligence	The identification, collection and use of internal and external data to support **eProcurement**.
eKeiretsu	Grouping of companies linked horizontally and vertically via the Internet to achieve a common objective.
eManufacturing	Internet- and intranet-enabled integration of manufacturing processes with each other and with the external supply chain.
eMarketplaces	Internet enabled groupings of companies aimed at sharing procurement costs and leveraging combined purchasing power.
Employee-centric portals	Single point of access to all tools, information and resources available to a company's employees.
Entraprise	Extended enterprise of trading partners.
ePayment	Payment for goods or services over the Internet.

eProcurement Electronic procurement of goods and services over the Internet via, for example, **eMarketplaces** or **eAuctions**.

eRequisitioning Automated order placement over the Internet, also referred to as **eBuying**.

ERP Enterprise resource planning.

eTailer Electronic – i.e. Web-based – retailer.

eVenture Any form of Internet/Web-based business.

eWorking Implementing Internet/intranet-enabled business processes.

Flow management Management of the flow of transactions through a supply chain network.

GPS Geographical positioning system.

Hardware platform The computer/electronic platform on which **applications** and system **software** run.

Holistic capability building decisions Strategy for development of **eProcurement** programmes based on the development of long-term online capabilities.

Horizontal eMarketplaces **eMarketplaces** in which the participants are drawn from different industrial and commercial sectors.

iCity Intelligent – i.e. Internet enabled – city.

Infomediaries Intermediate providers of information in a supply chain network to support synchronized decision-making.

Infrastructure platforms What used to be called operating systems, for example, Unix, Windows NT and Windows 2000.

Instant messaging	Internet-enabled interactive communication in real time (chat).
Inter-industry	The phenomenon of one industry's demand side being another industry's supply side.
IPO	Initial public offering.
IT	Information technology.
iTV	Interactive television.
JV	Joint venture.
Kanban processes	Processes supporting just-in-time replenishment of materials and components direct to manufacturing operations.
KPI	Key performance indicator.
Lead-time	Time between an order being placed and the product or goods being delivered.
Make-to-order	Manufacture of item commences only when order is placed; no stock of finished product.
Mass customization	Production of mass-produced items incorporating customer specified variations on a **make-to-order** basis.
mCommerce	Mobile commerce – **eBusiness** conducted through wireless communications devices, including **WAP** phones, **PDAs** etc.
Metrics	Measures of business performance.
Middleware	Specialist software which allows different **software applications**, often on different **hardware and infrastructure platforms**, to exchange data.
MRO	Maintenance, repair and operations.

MRP/MRP2	Manufacturing (or materials) resources planning.
OEM	Original equipment manufacturer.
Onboarding	The process of ensuring that a vendor's products are available through a particular eMarketplace.
Online collaboration	Collaboration between companies over the Internet, especially via an **eMarketplace**.
Order-to-cash	The complete business process cycle from receiving an order to supplying the goods and receiving payment for them.
Packaged interfaces	Sometimes called 'connectors' or 'adaptors' – standard software which reduces the development effort and bespoke code required to get information into and out of databases and **applications software**.
PDA	Personal digital assistant.
Platform	The hardware/operating system combination upon which **applications software** is supported.
Plug & Play	Self-configuring hardware and software – you plug it in, it plays.
Portal services	Services made available and delivered through a single common point of access.
Postponement manufacturing	Technique for reducing finished goods inventory.
R&D	Research and development.
Reverse auction	**eContracting** technique whereby potential purchaser offers business to lowest bidder.
RFI	Request for information.
RFP	Request for proposal.

RFQ Request for quotation.

Scalability The ability of a computer based hardware/software solution to expand to handle greater volumes of throughput without major discontinuities.

SCP Supply chain planning.

SKU Stock keeping unit.

Smartcard Credit/debit style card incorporating processing and memory capability.

SME Small to medium sized enterprise.

SMS Secure message service – method for sending encrypted messages over the Internet.

Start-ups New companies.

Strategic sourcing The identification of a strategic mix of suppliers to fulfill an organization's commodity and services demand with improved cost and service.

Systems application footprint That portion of the user's total requirement met by a particular system solution.

Systems integrator Specialist third party which brings together hardware and software from multiple vendors to provide a customer specific solution.

Track-and-trace Ability to monitor, report on and respond to queries about the movement of individual products or consignments throughout the distribution and delivery process.

Turnkey Complete systems solutions delivered, installed, configured and ready to operate 'at the turn of the key'.

Value chain The combination of supply chain and demand chain.

Virtual design environment	Computer-generated space within which engineers and designers can create, develop, test and optimize new product designs without recourse to the expense of real world facilities or prototypes.
Virtual reality	Computer generated, interactive modelling of real world facilities, functions and phenomena.
VMI	Vendor managed inventory.
WAP	Wireless application protocol.
Web browser tool	**Software application** implemented through a **browser-based user interface**.
Webcasting	Simultaneous, possibly interactive, real-time broadcast of text, audio and video to multiple recipients over the Internet.
Web conferencing	Multi-way, multi-participant real-time messaging over the Internet.

Index